AF839472

Urban Green Spaces and Environmental Sustainability

Md Babor Ali · Saleha Jamal

Urban Green Spaces and Environmental Sustainability

Insights from Kolkata Municipal Corporation

Springer

Md Babor Ali
Department of Geography
Faculty of Science
Aligarh Muslim University
Aligarh, Uttar Pradesh, India

Saleha Jamal
Department of Geography
Faculty of Science
Aligarh Muslim University
Aligarh, Uttar Pradesh, India

ISBN 978-3-032-15321-0 ISBN 978-3-032-15322-7 (eBook)
https://doi.org/10.1007/978-3-032-15322-7

© The Editor(s) (if applicable) and The Author(s), under exclusive license to Springer Nature Switzerland AG 2026

This work is subject to copyright. All rights are solely and exclusively licensed by the Publisher, whether the whole or part of the material is concerned, specifically the rights of translation, reprinting, reuse of illustrations, recitation, broadcasting, reproduction on microfilms or in any other physical way, and transmission or information storage and retrieval, electronic adaptation, computer software, or by similar or dissimilar methodology now known or hereafter developed.
The use of general descriptive names, registered names, trademarks, service marks, etc. in this publication does not imply, even in the absence of a specific statement, that such names are exempt from the relevant protective laws and regulations and therefore free for general use.
The publisher, the authors and the editors are safe to assume that the advice and information in this book are believed to be true and accurate at the date of publication. Neither the publisher nor the authors or the editors give a warranty, expressed or implied, with respect to the material contained herein or for any errors or omissions that may have been made. The publisher remains neutral with regard to jurisdictional claims in published maps and institutional affiliations.

This Springer imprint is published by the registered company Springer Nature Switzerland AG
The registered company address is: Gewerbestrasse 11, 6330 Cham, Switzerland

If disposing of this product, please recycle the paper.

Dedicated
To
Our Beloved Teachers
and Parents

Foreword

Urban green spaces have emerged as vital components of sustainable cities, serving as the lungs of the urban ecosystem and as essential public assets that enhance the quality of life, promote environmental balance, and foster social cohesion. Rapid urbanization and widespread urban sprawl have depleted green cover and increased urban vulnerability to climate change. Increase in population and their intense activities have adversely affected the green cover among metropolitan cities of India. Despite several initiatives, it remains far below than the UN recommended standard of 9 m^2 of green space per capita, making the preservation and enhancement of urban green spaces crucial for biodiversity.

The book *Urban Green Spaces and Environmental Sustainability: Insights from Kolkata Municipal Corporation* is both a wake-up call and a beacon of hope. The book highlights intricate relationship between green spaces, urban development, and environmental well-being in one of India's historical city of Kolkata, often referred to as the "Cultural Capital of India." The city from colonial-era architecture to modern urban development portrays a unique blend of tradition and progress. This work is both scholarly and policy-oriented, combining rigorous scientific analysis with practical insights. Through the integration of effective geospatial techniques, the authors effectively evaluate the spatial distribution, environmental significance, and vulnerability of urban green spaces within the Kolkata Municipal Corporation (KMC) area. The book not only maps patterns of change over three decades but also provides an in-depth understanding of how these transformations influence air quality, thermal regulation, biodiversity, and public health. The emphasis on the socio-environmental role of urban greenery, its contribution to psychological well-being, physical health, and climate resilience, invites readers to confront the uncomfortable truths about our environmental footprint while illuminating pathways toward sustainability, resilience, and renewal. Equally important are the author's policy recommendations, which is an invaluable contribution to contemporary urban studies.

This book will serve as a valuable reference for urban planners, geographers, environmental scientists, policymakers, and students interested in the intersection of urbanization and ecological sustainability. It provides a roadmap for reimagining cities as ecologically balanced and inclusive spaces, where development and nature coexist harmoniously.

I commend the authors for this significant contribution to the growing body of literature on urban environmental management and sustainability.

Prof. Shahab Fazal
Professor
Department of Geography
Aligarh Muslim University
Aligarh, India

Preface

The book *Urban Green Spaces and Environmental Sustainability: Insights from Kolkata Municipal Corporation* presents an in-depth analysis of the dynamic relationship between urbanization, environmental sustainability, and the role of green spaces in enhancing urban resilience. In the contemporary context of rapid urban expansion, cities across the globe and particularly in developing nations like India are facing significant environmental challenges, including loss of biodiversity, declining air quality, and rising urban heat island effects. Kolkata, one of India's oldest and most densely populated metropolitan cities, exemplifies these challenges, making it a compelling case for examining the status, vulnerability, and future of urban green spaces (UGS).

This book explores the spatial and ecological dimensions of UGS within the jurisdiction of the Kolkata Municipal Corporation (KMC), integrating geospatial analysis, field-based assessments, and socio-environmental perspectives. It systematically evaluates land use and land cover changes, environmental impacts, and the vital role of green spaces in mitigating pollution, regulating temperature, and improving psychological and physical well-being. The study employs advanced analytical tools such as Geographic Information Systems (GIS), remote sensing (RS), Analytic Hierarchy Process (AHP), and CA-Markov modeling to assess the current status, vulnerability, and projected future scenarios of urban green areas.

Moreover, the book emphasizes the importance of policy interventions and sustainable planning strategies to conserve and expand green infrastructure. It provides evidence-based insights and practical recommendations for urban planners, policymakers, researchers, and environmentalists to foster greener, healthier, and more sustainable cities.

The work represents a comprehensive attempt to bridge the gap between environmental research and practical urban policy, highlighting that sustainable urban futures depend not merely on technological advancements but on a deeper commitment to preserving and revitalizing nature within the urban fabric.

This book represents a revised and expanded edition of the doctoral dissertation submitted to Aligarh Muslim University, with its content extensively rewritten and updated to address the expectations of a wider academic and professional readership.

Aligarh, India

Dr. Md Babor Ali
Dr. Saleha Jamal

Acknowledgements The authors would like to express their sincere gratitude to the Department of Geography and the Maulana Azad Library, Aligarh Muslim University, Aligarh, for providing institutional support and access to essential academic resources that facilitated the successful completion of this work. The authors are also thankful to the University Grants Commission (UGC) for extending financial assistance during the Ph.D. tenure, which substantially aided the research process.

The authors further acknowledge the Kolkata Municipal Corporation (KMC), Kolkata Metropolitan Development Authority (KMDA), Kolkata Police, and the Kolkata Pollution Control Board (KPCB) for providing the requisite permissions, datasets, and technical inputs necessary for the empirical analysis presented in this book. The scholarly contributions of various researchers and academicians, whose work provided the conceptual and methodological underpinnings of this study, are also gratefully acknowledged.

Thanks are also due to all individuals who have supported this endeavor, directly or indirectly, and whose encouragement and goodwill have been instrumental in the completion of this work.

Aligarh, India

Dr. Md Babor Ali
Dr. Saleha Jamal

Competing Interests The authors have no competing interests to declare that are relevant to the content of this manuscript.

Contents

About the Authors

Dr. Md Babor Ali is a Research Associate in the Department of Geography at Aligarh Muslim University, specializing in urban ecology, geoinformatics, and environmental sustainability. He holds a Ph.D. in Geography from AMU, where his doctoral research focused on urban ecology, urban green spaces, and urban sustainability. A UGC-JRF/SRF awardee, he has published extensively in high-impact international journals (Springer, Taylor & Francis) and serves as a reviewer for several leading Q1 journals. Proficient in GIS and Remote Sensing applications, Dr. Ali integrates advanced geospatial modeling with sustainability research, contributing significantly to policy-oriented environmental planning and urban resilience studies.

Dr. Saleha Jamal is an Associate Professor in the Department of Geography at Aligarh Muslim University (AMU), with expertise in urban environment, environmental health, geospatial applications, and sustainability studies. A recipient of numerous national awards, including the "Best Geography Teacher Award 2023" and several prestigious young geographer awards, she has published extensively in reputed international journals and edited volumes. She has authored and co-edited multiple books with Springer and Taylor & Francis Group and has successfully completed ICSSR-funded research projects. As a dedicated academician and mentor, she has supervised doctoral and postdoctoral research on urban sustainability, health, and environmental governance.

Abbreviations

AHP	Analytic Hierarchy Process
AOI	Area of Interest
AQI	Air Quality Index
CCC	Centre for Contemporary Communication
CPCB	Central Pollution Control Board
CR	Consistency Ratio
DEM	Digital Elevation Model
DN	Digital Number
EKW	East Kolkata Wetlands
EKWA	East Kolkata Wetlands Act
FCC	False Color Composites
GCP	Ground Control Points
GIS	Geographic Information System
GPS	Global Positioning System
GSV	Green Space Vulnerability
IDW	Inverse Distance Weighting
KMC	Kolkata Municipal Corporation
KMDA	Kolkata Metropolitan Development Authority
KPCB	Kolkata Pollution Control Board
LSG	Local Self-Governing
LST	Land Surface Temperature
LULC	Land Use Land Cover
MLC	Maximum Likelihood Classification
MLP	Multi-Layer Perceptron
MSL	Mean Sea Level
NCR	National Capital Region
NDBI	Normalized Difference Built-up Index
NDVI	Normalized Difference Vegetation Index
NGO	Non-Governmental Organization
NIR	Near Infrared
OA	Overall Accuracy

OLI	Operational Land Imager
OSM	OpenStreetMap
PM	Particulate Matter
PSR	Pressure State Response
PWD	Public Works Department
RRSC	Regional Remote Sensing Centre
SPSS	Social Science Statistical Package
SSKM	Seth Sukhlal Karnani Memorial
SWIR	Shortwave Infrared
TIRS	Thermal Infrared Sensor
TM	Thematic Mapper
UGS	Urban Green Space
UHI	Urban Heat Island
USGS	United States Geological Survey
UTM	Universal Transverse Mercator
VCI	Vegetation Condition Index
WBR	World Bank Report
WGS	World Geodetic System
WHO	World Health Organization
WO	Weighted Overlay

List of Figures

List of Tables

Chapter 1
Introduction

1.1 Overview of Urban Green Spaces and Its Importance in Environmental Sustainability

Urban green spaces (UGS) constitute an integral component of the urban ecosystem, contributing significantly to environmental quality, human well-being, and the overall sustainability of cities. They include parks, gardens, street trees, playgrounds, wetlands, green roofs, and other vegetated areas within the built environment. These spaces act as the "green lungs" of cities by maintaining ecological balance, improving air quality, reducing the urban heat island effect, and supporting biodiversity. In the context of rapidly urbanizing regions, particularly in developing countries, the importance of green spaces extends beyond aesthetic or recreational value to encompass critical environmental, social, and economic functions that are essential for sustainable urban living.

The rapid pace of urbanization has resulted in extensive land use transformations, leading to the loss of natural habitats and open spaces. As cities expand, green areas are frequently replaced by impermeable surfaces such as concrete, asphalt, and glass, which disrupt natural processes like water infiltration, evapotranspiration, and temperature regulation. This transformation has exacerbated environmental problems, including rising surface temperatures, deteriorating air and water quality, and increased vulnerability to flooding. Urban green spaces play a vital role in mitigating these adverse impacts by performing ecosystem services such as air purification, carbon sequestration, and microclimatic regulation. Vegetation absorbs atmospheric carbon dioxide and other pollutants while releasing oxygen, thus reducing greenhouse gas concentrations and improving urban air quality.

From an ecological perspective, UGS contribute to maintaining biodiversity within cities by providing habitats, food sources, and migratory corridors for a variety of plant and animal species. Even small patches of urban greenery, such as street trees or community gardens, can create ecological niches that enhance species diversity.

© The Author(s), under exclusive license to Springer Nature Switzerland AG 2026
M. B. Ali and S. Jamal, *Urban Green Spaces and Environmental Sustainability*,
https://doi.org/10.1007/978-3-032-15322-7_1

These spaces also contribute to soil stability, groundwater recharge, and the prevention of erosion. Furthermore, urban wetlands and green belts act as natural buffers that reduce flood risk and maintain hydrological cycles, thereby increasing the city's resilience to climate change.

The social and psychological benefits of green spaces are equally significant. Numerous studies have shown that exposure to nature enhances mental health, reduces stress, and improves overall well-being. Green spaces provide venues for recreation, physical exercise, and social interaction, fostering a sense of community and inclusivity among urban residents. They also play an important role in reducing social inequalities by offering accessible natural environments for all sections of society, regardless of income or social status. In densely populated urban settings, public parks and gardens offer essential relief from congestion, noise, and pollution, improving the quality of urban life.

Economically, green spaces contribute to the value of urban properties, attract investment, and stimulate local economies through recreation, tourism, and environmental services. Cities that integrate green infrastructure into their planning frameworks tend to demonstrate higher resilience to environmental stresses and improved living standards. The inclusion of UGS in urban planning aligns with the principles of Sustainable Development Goal 11 (SDG 11), which emphasizes the creation of inclusive, safe, resilient, and sustainable cities.

In summary, urban green spaces are vital assets for promoting environmental sustainability in metropolitan regions. They serve as multifunctional systems that integrate ecological, social, and economic processes within urban landscapes. By conserving and expanding these spaces, city planners can enhance environmental quality, strengthen climate resilience, and ensure a healthier and more sustainable urban future. The sustainability of modern cities, therefore, depends greatly on how effectively they protect, manage, and integrate urban green spaces within their developmental agendas.

1.2 Defining Key Concepts: Urban Ecology, Green Infrastructure and Ecosystem Services

Understanding the environmental significance of urban green spaces requires conceptual clarity regarding three interrelated frameworks—urban ecology, green infrastructure, and ecosystem services. Together, these concepts form the theoretical foundation for assessing how natural and semi-natural systems function within the built environment and contribute to urban sustainability.

Urban ecology is an interdisciplinary field that studies the interactions between living organisms and their urban environments, combining elements of ecology, geography, sociology, and urban planning. Unlike traditional ecology, which primarily examines natural ecosystems, urban ecology focuses on human-dominated landscapes where natural and artificial systems coexist and influence one another.

The core principle of urban ecology is that cities are dynamic ecosystems with complex interdependencies among biophysical, social, and economic components. Human activities such as land use change, industrialization, and infrastructure development alter ecological processes, including nutrient cycling, hydrology, and biodiversity patterns. However, through sustainable design and planning, urban systems can be restructured to mimic natural processes and improve environmental quality. Urban ecology thus provides a scientific framework for understanding how urbanization affects ecological balance and how green spaces can restore or maintain this equilibrium.

Green infrastructure (GI) is another key concept that operationalizes the ideas of urban ecology by integrating ecological principles into spatial planning and design. It refers to a strategically planned network of natural and semi-natural spaces such as parks, forests, wetlands, green roofs, and urban gardens that collectively provide environmental, social, and economic benefits. Green infrastructure is not limited to individual green patches but emphasizes connectivity among them, forming an interconnected system that supports biodiversity, regulates microclimates, and enhances urban resilience. Unlike traditional "gray infrastructure," which relies on engineered systems like pipes and concrete for urban services, GI uses nature-based solutions to deliver the same or even superior outcomes at lower environmental and economic costs. Examples include permeable pavements that facilitate water infiltration, urban forests that mitigate air pollution, and wetlands that manage stormwater. The European Commission (2013) defines green infrastructure as "a strategically planned network of natural and semi-natural areas designed and managed to deliver a wide range of ecosystem services." This concept highlights the multifunctionality of green spaces and their potential to support sustainable urban development by aligning environmental conservation with socio-economic objectives.

Ecosystem services are the direct and indirect benefits that humans derive from natural ecosystems, encompassing provisioning, regulating, cultural, and supporting services. In urban contexts, these services are critical for maintaining environmental stability and enhancing human well-being. Provisioning services include tangible products such as food, water, timber, and medicinal resources. Regulating services involve natural processes like air purification, climate regulation, flood control, and carbon sequestration. Cultural services refer to the recreational, aesthetic, spiritual, and educational values provided by green spaces, while supporting services maintain the ecological foundations that make all other services possible such as soil formation, nutrient cycling, and biodiversity maintenance. Urban green spaces, by offering these multiple functions, act as essential providers of ecosystem services that underpin the sustainability of urban systems.

In essence, urban ecology provides the theoretical foundation for understanding cities as ecosystems; green infrastructure offers a practical framework for incorporating ecological systems into urban planning; and ecosystem services represent the measurable benefits derived from these ecological processes. The integration of these three concepts forms the cornerstone of contemporary sustainable urban development. Together, they emphasize that cities are not separate from nature but are deeply embedded within it. Building resilient, livable, and sustainable urban environments, therefore, depends on how effectively urban planners, policymakers, and communities integrate ecological thinking into the design and governance of urban spaces.

1.3 Background and Rationale for Selecting Kolkata Municipal Corporation as the Study Area

Kolkata Municipal Corporation (KMC) has been chosen as the focal area for this study due to its unique historical evolution, demographic density, and ecological vulnerability. As one of the oldest municipal bodies in India, KMC represents the core of the Kolkata Metropolitan Area, encompassing a complex interplay of rapid urbanization, socio-economic diversity, and environmental stress. The city's urban fabric has undergone extensive transformation since the colonial period, evolving from a riverine port settlement into a megacity characterized by intense land use pressures, industrial growth, and population concentration. These factors collectively make Kolkata an ideal case for assessing the environmental impacts of urban green spaces within a highly urbanized tropical setting.

Kolkata's geography situated on the eastern bank of the Hooghly River in the lower Gangetic delta, renders it particularly sensitive to climatic and hydrological challenges such as flooding, drainage congestion, and rising temperatures. The city's flat terrain, dense population, and extensive built-up area exacerbate these problems, resulting in significant ecological imbalance. The reduction of natural vegetation and open spaces has intensified issues like air pollution, heat stress, and waterlogging, making environmental management a critical urban concern.

KMC's administrative area is characterized by uneven distribution of green spaces across its 144 wards and 16 boroughs, reflecting sharp socio-spatial disparities in environmental quality and accessibility. These variations offer a valuable framework for evaluating the interrelationship between urban morphology, ecological degradation, and public well-being. Furthermore, Kolkata's position as a major economic and cultural hub in eastern India underscores the necessity of integrating environmental sustainability into its urban development agenda.

Therefore, selecting KMC as the study area allows for a comprehensive investigation into how urban green spaces contribute to environmental quality, resilience, and sustainability, providing insights applicable to other fast-growing cities in the Global South.

1.4 Statement of the Problem

Urban green spaces (UGS) play a crucial role in maintaining ecological balance, enhancing environmental quality, and promoting human well-being within cities. However, rapid urbanization, population growth, and the expansion of built-up areas have placed tremendous pressure on these essential natural assets. Across most developing nations, particularly in India, the pursuit of economic growth and infrastructural development has often overshadowed the importance of preserving urban greenery. Consequently, the natural landscape within cities is shrinking at an alarming rate, leading to deteriorating environmental health, loss of biodiversity, and declining quality of life for residents.

In the case of Kolkata, one of India's oldest and most densely populated metropolitan cities, this problem is especially pronounced. The city has undergone extensive spatial and demographic transformation, resulting in the conversion of open and green spaces into residential, commercial, and industrial land uses. This shift has led to significant changes in land use and land cover (LULC) patterns, contributing to issues such as increased Land Surface Temperature, poor air quality, urban flooding, and reduced environmental resilience. Despite the numerous environmental and social benefits of green spaces including carbon sequestration, air purification, groundwater recharge, and recreational functions, these areas continue to be undervalued and under threat from urban expansion.

Another major concern lies in the uneven spatial distribution and accessibility of urban green spaces across the city. While central and affluent neighborhoods of Kolkata may still retain some level of greenery, peripheral and densely populated areas experience a severe deficiency. This unequal distribution results in environmental injustice and limits the social, psychological, and health benefits that green spaces provide. The problem is further compounded by weak urban governance, inadequate enforcement of planning regulations, and insufficient public participation in environmental management.

Given these challenges, there is an urgent need to assess the environmental impacts of urban green spaces within the jurisdiction of the Kolkata Municipal Corporation (KMC). Such an assessment must focus not only on quantifying the current state and spatial distribution of green areas but also on understanding their ecological functions and vulnerability to urban pressures. A scientifically informed evaluation of urban green spaces can provide valuable insights into sustainable urban planning and environmental policy formulation. It can also help strengthen urban resilience and contribute to achieving broader sustainability goals in line with global frameworks such as the Sustainable Development Goals (SDGs).

Therefore, this research seeks to address a critical gap by examining the environmental implications of changing green space patterns in Kolkata, assessing their contribution to urban sustainability, and identifying areas most vulnerable to environmental degradation. Through the integration of geospatial analysis, environmental indicators, and field-based observations, this study aims to provide a comprehensive understanding of the relationship between urbanization and the city's green infrastructure, thereby contributing to more informed decision-making for sustainable urban development.

1.5 Conceptual Framework

The conceptual framework of this research is grounded in the interdisciplinary relationship between urban ecology, environmental sustainability, and geospatial analysis. It seeks to explore how urban green spaces (UGS) influence the environmental quality and ecological resilience of a metropolitan city like Kolkata, where continuous land transformation and population pressure have disrupted the natural equilibrium of the urban ecosystem. The framework integrates multiple dimensions e.g. ecological, spatial, and socio-environmental to provide a holistic understanding of the role and dynamics of UGS within the broader context of urban sustainability.

At its core, the study conceptualizes urban green spaces as essential components of the urban ecosystem that interact with various physical, biological, and human processes. These interactions affect the city's environmental conditions, such as air quality, temperature regulation, and biodiversity preservation. The conceptual model posits that the spatial extent, distribution, and connectivity of green spaces are critical determinants of ecological balance and environmental health in urban settings (Tzoulas et al., 2007). Green spaces not only serve as carbon sinks but also mitigate the adverse effects of rapid urbanization, including the urban heat island effect, noise pollution, and deteriorating air quality.

The framework also emphasizes the Pressure-State-Response (PSR) approach as a guiding analytical model for assessing the environmental impacts of UGS. This model recognizes that urban systems are dynamic and continuously influenced by anthropogenic pressures. "Pressure" refers to the factors exerted by human activities such as population density, industrialization, and land use changes that contribute to environmental degradation. "State" represents the existing condition of the environment, including vegetation cover, air and water quality, and spatial distribution of green areas. "Response" pertains to the policy measures, planning strategies, and community initiatives designed to mitigate pressures and restore environmental balance. Through this PSR model, the study aims to establish a structured and quantifiable linkage between human-induced pressures and the ecological state of green spaces within Kolkata.

Furthermore, the conceptual framework incorporates spatial modeling and geoinformatics techniques such as Analytic Hierarchy Process (AHP) and Cellular Automata-Markov (CA-Markov) models to evaluate the vulnerability and future trajectories of UGS. The AHP model facilitates the weighting of various environmental and spatial indicators based on their relative importance, allowing for a systematic prioritization of factors affecting green space vulnerability. Meanwhile, the CA-Markov model supports the simulation of land use and land cover (LULC) changes, providing valuable insights into the possible future scenarios of urban green spaces. Together, these methods enable a multi-dimensional and predictive assessment of UGS sustainability under changing urban conditions.

Another key element of this conceptual framework lies in understanding the human–environment relationship. Urban green spaces are not merely ecological entities but also social infrastructures that enhance community well-being, promote recreational opportunities, and foster environmental awareness among citizens. The framework recognizes that sustainable urban development cannot be achieved without integrating social perceptions, local knowledge, and participatory governance mechanisms. Hence, the conceptual structure bridges the gap between environmental science, urban planning, and social sustainability.

In summary, the conceptual framework of this research establishes a comprehensive analytical structure to examine urban green spaces in Kolkata through an integrated perspective of ecological function, spatial dynamics, and socio-environmental interaction. It provides the foundation for evaluating the environmental impacts of UGS and supports the formulation of policy recommendations aimed at promoting sustainable urban growth while preserving the city's ecological integrity.

1.6 Objectives of the Study

The study aims to achieve the following objectives.

- To assess the transformation status, evaluation, and distribution of green space.
- To assess the role of urban green spaces on thermal variability; and relationship between green spaces and surface temperature.
- To assess the concentration of different parameters of air quality at different stations and their association with green spaces.
- To examine the noise level at different stations and its association with green spaces.
- To assess the impacts of urban green spaces on human health (psycho-social health and physical health).
- To make the identification and evaluation of the vulnerable zones of urban green spaces.
- To suggest suitable measures and strategies for conservation, enhancement and Policy Recommendations of urban green spaces in KMC.

1.7 Research Questions

The current study will attempt to answer the following questions.

- Does the land use land cover transformation have impacted the urban green spaces in Kolkata Municipal Corporation?
- Does urban green spaces regulate the microclimate in the city?

- Does there is role of urban green spaces on environmental pollution abatement in the study area?
- Is there any association between urban green spaces and resident's health?

1.8 Hypotheses

Following are the hypotheses that will be tested based on primary data.

- There is significant effect of availability of green space on resident's psycho-social health.
- There is significant effect of availability of green spaces on resident's physical health.

1.9 Database and Methodology

For this research, both primary and secondary sources of data have been used and their sources are given briefly below.

Primary data.

- Schedule method.
- Interview method.
- Field survey.
- GPS.

Secondary Data.

- Landsat-8 (Operational Land Imager) for the year 2020 (USGS).
- Landsat-5 (Thematic Mapper) for the years 1990, 2000, and 2010 (USGS).
- Central Pollution Control Board.
- Kolkata Pollution Control Board.
- Kolkata Municipal Corporation Office.
- Kolkata Municipal Development Authority.
- Google Earth Pro.
- Topographical Map.
- Open Steet Map Data.
- Census of India Data.

The methodology adopted in this research integrates both spatial and non-spatial data to examine land use and land cover (LULC) transformations, with particular reference to urban green spaces (UGS) within the jurisdiction of the Kolkata Municipal Corporation (KMC). The spatial data primarily comprise multi-temporal satellite imagery from Landsat-5 Thematic Mapper (TM) and Landsat-8 Operational Land Imager (OLI), covering the period between 1990 and 2020, sourced from the USGS Earth Explorer platform. Complementing these, non-spatial datasets were collected from the Census of India and relevant secondary government reports. To ensure reliability and accuracy, extensive ground-truthing was performed using Global Positioning System (GPS) technology, generating 127 GPS reference points and 108 Google Earth-based validation points.

Supervised classification of satellite imagery was carried out using the Maximum Likelihood Classification (MLC) algorithm. An Area of Interest (AOI) was delineated from each image, and 100 training samples per land use class were selected for classification. These were cross-verified with Google Earth imagery and topographical data to ensure classification accuracy. The process was repeated for four temporal datasets i.e. 1990, 2000, 2010, and 2020, producing a series of LULC maps that illustrate the city's changing landscape. The classification and analysis were conducted using ERDAS IMAGINE 2014 and ArcMap 10.5 software. Statistical analyses, including linear regression, were used to evaluate the relationship between population growth, built-up expansion, and the reduction of green spaces. Scatter plots and regression equations were developed to visualize trends. Cohen's Kappa coefficient was applied to measure classification accuracy and validate the reliability of results.

In the third chapter, the methodology focuses on analyzing the city's environmental conditions through indicators such as the Normalized Difference Built-up Index (NDBI), Normalized Difference Vegetation Index (NDVI), and Land Surface Temperature (LST) for 1990, 2000, 2010, and 2020. Temporal comparisons were made at decadal intervals to assess the evolution of urban heat and vegetation loss. A borough-level analysis was performed using ArcGIS 10.5, employing clip tools to extract spatial data on built-up area, vegetation index, and surface temperature. Statistical evaluations in SPSS software were utilized to examine correlations between LST, urbanization, and green cover. Regression models and scatter plots were used to establish and visualize the relationships among these environmental variables.

The fourth chapter adopts a detailed approach to assess air quality and noise pollution within the study area. Seven air quality monitoring stations were strategically selected based on their proximity to UGS, while nine noise level monitoring sites were distributed across areas with varying degrees of vegetation cover. Air quality parameters carbon monoxide (CO), nitrogen oxide (NO), sulfur dioxide (SO_2), and particulate matter ($PM_{2.5}$ and PM_{10}) were systematically recorded using standardized instruments and procedures. The data generated were essential for understanding the environmental impacts of urbanization and for developing recommendations for air pollution control and public health management. The chapter combines geospatial interpolation techniques with field-based monitoring to present a robust methodology for evaluating the ecological and health implications of environmental pollution in urban settings.

The fifth chapter focuses on the relationship between urban green spaces and human health, using both psychological and physical well-being indicators. A structured interview-cum-schedule technique was adopted for primary data collection, balancing the advantages of interviews and questionnaires while minimizing interviewer bias. A random sampling method ensured proportional representation of the population. The Likert Scale was used to measure residents' attitudes and perceptions toward UGS and their psychological benefits. Statistical tools such as Karl Pearson's Coefficient of Correlation were employed to examine the association between UGS and psycho-physical health outcomes, while paired two-tailed t-tests were used to compare related datasets and determine significant differences. The hypotheses were formulated to evaluate the effects of UGS availability on the physical and mental health of residents, forming a comprehensive methodology for linking ecological and social dimensions.

The sixth chapter analyzes the vulnerability of urban green spaces using both spatial and non-spatial data obtained from sources such as OpenStreetMap, Landsat imagery, topographical sheets, and Google Earth. The methodology is built upon the Pressure-State-Response (PSR) framework to systematically evaluate the drivers of UGS vulnerability. Indicators were identified and weighted through expert consultations, particularly with specialists in environmental science and ecology. The Analytic Hierarchy Process (AHP) was used to assign relative weights to indicators, facilitating a multi-criteria decision-making approach. These indicators were integrated using the Weighted Overlay (WO) technique to develop a green space vulnerability (GSV) map, highlighting areas of environmental stress. Additionally, advanced modeling tools such as the Multi-Layer Perceptron (MLP) and CA-Markov chain models were employed to simulate future LULC changes and forecast UGS scenarios based on variables such as proximity to built-up zones, transportation networks, elevation, and slope. The accuracy of the classified maps was validated through ground-truth data and Kappa coefficient analysis, ensuring scientific rigor.

In sum, the study integrates a diverse set of spatial, statistical, and empirical methods to examine urban green space dynamics, environmental quality, and human well-being within Kolkata. The database and methodological framework outlined here provides a foundation for understanding urban ecological processes and guiding evidence-based planning for sustainable urban futures. Detailed methodological steps for each analytical component are discussed in the respective chapters.

1.10 Services of Urban Green Spaces

1.10.1 Environmental Services

1.10.1.1 Ecological Services

Urban green spaces provide a diverse range of ecological functions essential for maintaining environmental equilibrium in metropolitan regions. These include biodiversity conservation, air purification, microclimate regulation, and hydrological balance. Climatic parameters such as rainfall distribution, solar radiation, temperature, and humidity differ significantly in urban environments compared to rural settings. According to Heidt and Neef (2008), urban areas experience altered climatic conditions owing to dense built-up structures that modify heat retention and airflow patterns. The accumulation of heat by impervious surfaces and the intensive energy consumption typical of metropolitan regions contribute to the *urban heat island* (*UHI*) effect, which can elevate city temperatures by up to 5 °C (Bolund & Hunhammar, 1999). Strategically distributed and adequately maintained urban green spaces play a critical role in moderating this phenomenon by providing shade, facilitating evapotranspiration, and improving air circulation, thereby reducing localized temperature extremes.

1.10.1.2 Pollution Control

Urbanization and industrialization have led to increased concentrations of air and noise pollutants in metropolitan areas, severely impacting public health and environmental quality. Major urban pollutants include particulate matter, carbon monoxide, sulfur dioxide, nitrogen oxides, and volatile organic compounds emitted from vehicles, factories, and other anthropogenic activities. These pollutants adversely affect human health, particularly among children, the elderly, and individuals with respiratory illnesses (Sorensen et al., 1997). Vegetation acts as a natural filter leaves and canopies capture dust, absorb gaseous pollutants, and reduce atmospheric contamination. Studies have shown that well-maintained urban parks can remove up to 85% of airborne pollutants (Bolund & Hunhammar, 1999).

Noise pollution, another byproduct of urban life, results from vehicular traffic, construction, and industrial activity. Its health consequences include stress, sleep disorders, and hearing impairment. Bolund and Hunhammar (1999) estimated that noise-related economic losses account for approximately 0.2–2% of the European Union's GDP. Vegetation buffers such as dense tree belts and parklands can effectively reduce ambient noise levels depending on their density, type, and distance from noise sources. In rapidly urbanizing nations like India and China, the conservation and expansion of urban greenery are indispensable for environmental management. Sustaining green infrastructure not only mitigates pollution but also supports carbon absorption, thereby contributing to global climate change mitigation (Huang et al., 2009).

1.10.1.3 Biodiversity and Nature Conservation

Urban green spaces act as crucial habitats for diverse flora and fauna, providing breeding grounds and ecological corridors that support urban biodiversity. These areas facilitate the interaction between urban and peri-urban ecosystems, contributing to soil and water conservation while enhancing landscape aesthetics. As Marcus and Francis (1997) highlight, green spaces create opportunities for residents to connect with nature, observe seasonal variations, and enjoy aesthetic relief amidst urban density. Loures et al. (2007) further emphasizes that integrating native vegetation and sustainable landscaping practices into city design ensures low maintenance costs and ecological self-sufficiency, forming the foundation for environmentally balanced urban morphology.

1.10.1.4 Provision of Water Drainage Systems

Urban green spaces significantly contribute to hydrological regulation by facilitating rainwater infiltration, replenishing groundwater, and reducing surface runoff. Impervious surfaces in cities often exacerbate flooding and groundwater depletion; however, vegetated areas help mitigate these effects. In Stockholm, Sweden, for instance, integrating green infrastructure within stormwater systems reduced maintenance costs and improved drainage efficiency (Bolund & Hunhammar, 1999). Similarly, Bryant (2006) also observed that integrating green drainage networks can function as both stormwater management tools and biodiversity buffers, protecting aquatic habitats from urban encroachment.

1.10.1.5 Provision of Sewage Treatment

Wetlands and vegetated landscapes play a critical role in wastewater treatment by absorbing and filtering excess nutrients such as nitrogen and phosphorus. These processes reduce eutrophication and improve overall water quality (Bolund & Hunhammar, 1999). Vymazal (2007) found that vegetation within constructed wetlands efficiently removes nitrogen, phosphorus, and heavy metals from wastewater, making such systems particularly effective in densely industrialized urban environments. The integration of diverse wetland species and configurations enhances nutrient removal potential, demonstrating that urban green infrastructure can serve as an eco-friendly and cost-effective alternative to conventional sewage treatment methods.

1.10.2 Economic and Aesthetic Services

1.10.2.1 Energy Savings

Urban vegetation significantly contributes to energy conservation by regulating air temperature and reducing the demand for artificial cooling. Trees and plants provide shade, facilitate evapotranspiration, and moderate air circulation, collectively reducing heat absorption by buildings (Heidt & Neef, 2008). A study conducted in Chicago revealed that expanding green cover by 10% could reduce overall heating and cooling energy consumption by 5–10% (Sorensen et al., 1997). Large urban parks also have localized cooling effects; for example, a park measuring 1.2 km by 1.0 km can influence air temperatures within a 4 km radius (Heidt & Neef, 2008).

Such natural cooling mechanisms make urban greenery a cost-effective approach to sustainable energy management in metropolitan regions.

1.10.2.2 Property Value

The presence of urban greenery enhances the aesthetic and economic value of surrounding properties. Landscaped areas and tree-lined streets increase residential appeal, attracting both residents and investors. Cities such as Singapore and Kuala Lumpur have successfully leveraged their green cityscapes to draw international investment and stimulate economic growth (Sorensen et al., 1997). Empirical studies show that properties located near well-maintained green spaces can yield financial returns between 5 and 15%, depending on the project and location (Heidt & Neef, 2008). Thus, green infrastructure not only contributes to environmental sustainability but also serves as an important driver of urban economic vitality.

1.10.3 Socio-psychological Services

1.10.3.1 Recreation and Well-Being

Urban green spaces are vital for recreation, leisure, and mental restoration. A study in 2000 reported that over 80% of the UK's population resides in urban areas, where access to local parks and open spaces fulfills a significant portion of their outdoor recreational needs. Similarly, another study in Helsinki found that nearly 97% of the population engages in outdoor activities annually, with over half visiting green spaces daily or every other day (Neuvonen et al., 2007). Green spaces provide residents with opportunities for relaxation, social interaction, and emotional rejuvenation (Heidt & Neef, 2008). In Mexico City, the Chapultepec Park attracts approximately three million visitors each week, underscoring the social and recreational significance of urban parks (Sorensen et al., 1997).

1.10.3.2 Human Health

Exposure to natural environments has well-documented physiological and psychological benefits. Individuals living in greener neighborhoods exhibit lower stress levels compared to those in built-up urban areas (Bolund & Hunhammar, 1999). Patients with hospital rooms overlooking green landscapes recover faster by nearly 10%, and require significantly less pain medication than those facing non-green surroundings. Similarly, Grahn and Stigsdotter (2003) observed that residents who spend more time in urban green spaces report lower stress and improved mental well-being. Cleaner air, reduced noise, and enhanced environmental quality contribute to better physical health, particularly by reducing respiratory ailments. Sorensen et al.

(1997) emphasized that maintaining a close connection between urban residents and the natural environment enhances daily productivity, emotional stability, and overall quality of life.

1.10.4 Carbon Sequestration Potential of Urban Green Spaces

Urban green spaces are indispensable in mitigating climate change impacts due to their carbon sequestration capacity. Parks, gardens, and urban forests act as carbon sinks, absorbing atmospheric CO_2 through photosynthesis and storing it within biomass and soil. The vegetation comprising trees, shrubs, and grasses actively removes CO_2 from the air, releasing oxygen in return, thereby improving urban air quality. Expanding urban greenery enhances this sequestration capacity, contributing to significant reductions in greenhouse gas emissions. Beyond carbon storage, these spaces also help mitigate the *urban heat island* effect, foster biodiversity, and improve overall environmental resilience. Recognizing the carbon sequestration potential of UGS allows city planners to integrate green infrastructure into climate adaptation strategies, ensuring healthier, more sustainable, and climate-resilient cities.

1.11 Geographical Profile of the Study Area and Its Physical Settings

Kolkata, formerly known as Calcutta, serves as the capital of the Indian state of West Bengal and stands as the third-largest megacity in India. The city developed along the western bank of the Hooghly River, the principal distributary of the Ganga River. For the purpose of this research, the Kolkata Municipal Corporation (KMC) has been selected as the study area. It represents one of the four municipal corporations that collectively form the Kolkata Megacity. KMC constitutes the historical and administrative core from which the greater Kolkata region gradually expanded over time. The municipal jurisdiction is subdivided into 16 boroughs and further into 144 wards (Fig. 1.1). According to the 2011 Census of India, KMC accommodates a residential population of approximately 4.5 million and a floating population of nearly 6 million, with a population density of about 24,252 persons per square kilometer. Since its establishment, Kolkata has evolved into a primary center of industry, commerce, trade, and services, functioning as one of India's foremost urban and economic hubs (Dinda et al., 2021). Geographically, the city can be divided into five major zones north, south, east, west, and central Kolkata.

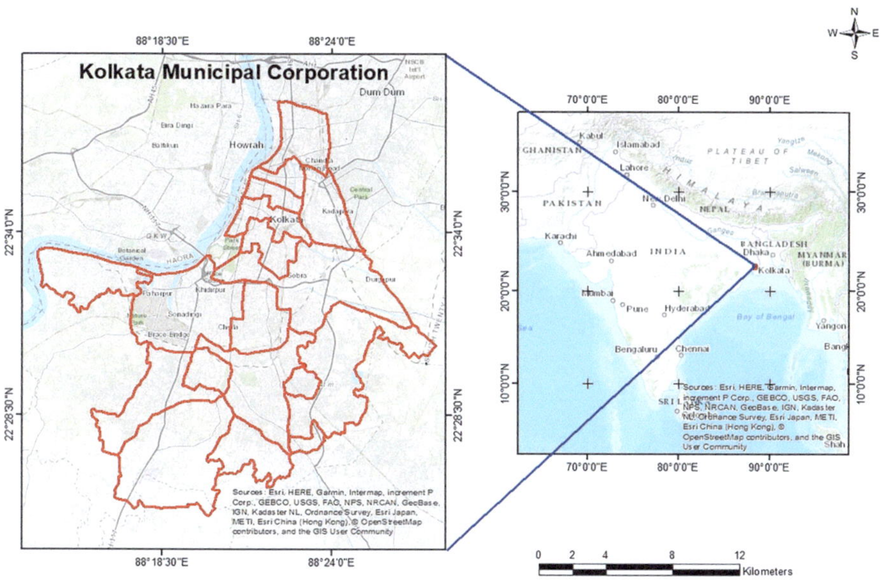

Fig. 1.1 Geographical location of the study area

1.11.1 Location and Boundaries

The Kolkata Municipal Corporation lies at approximately 22.5726°N latitude and 88.3639°E longitude, positioned within the Gangetic alluvial plains about 80 km west of the Bay of Bengal. It is bounded by the North 24 Parganas district to the north, South 24 Parganas to the south, Hooghly district to the west, and the Hooghly River to the east. This strategic location has historically facilitated trade, transport, and socio-economic integration with the broader Bengal deltaic region.

1.11.2 Topography

The topography of Kolkata is predominantly flat, gently sloping toward the east and southeast along the riverine system. The elevation of the city ranges between 1 and 9 m above mean sea level, making it highly susceptible to waterlogging and periodic flooding. The terrain is composed largely of alluvial deposits brought by the Hooghly River and its tributaries, which have shaped the city's geomorphological and hydrological characteristics over centuries.

1.11.3 Rivers and Water Bodies

The Hooghly River, a vital distributary of the Ganges River, flows along the eastern periphery of the city and divides the urban agglomeration into two parts, Kolkata on the east bank and Howrah district on the west. The river has historically supported navigation, commerce, and industry. Besides the Hooghly, the city features an intricate network of canals, ponds, and wetlands, which function as drainage and flood-mitigation systems. These water bodies not only support local biodiversity but also contribute to urban ecological balance and microclimatic regulation.

1.11.4 Climate

According to the Köppen–Geiger climate classification, Kolkata experiences a tropical wet and dry climate, characterized by distinct summer, monsoon, autumn, and winter seasons. Summers, extending from March to June, are hot and humid, with temperatures frequently exceeding 40 °C. The monsoon season (June to September) brings substantial rainfall under the influence of southwest monsoon winds, accounting for the majority of annual precipitation. Winters, from November to February, are mild and pleasant, with temperatures varying between 12 and 23 °C, offering comfortable climatic conditions. This distinct seasonal variation significantly influences vegetation growth, green cover distribution, and environmental processes across the city.

1.11.5 Green Space

Urbanization has considerably altered the natural vegetation of Kolkata and its adjoining areas. However, several prominent parks, gardens, and tree-lined avenues still contribute to the city's green character. Common tree species include mango (*Mangifera indica*), coconut (*Cocos nucifera*), neem (*Azadirachta indica*), banyan (*Ficus benghalensis*), and peepal (*Ficus religiosa*). The Acharya Jagadish Chandra Bose Botanical Garden, located in nearby Howrah, is one of the oldest and most renowned botanical gardens in Asia, housing an extensive collection of native and exotic flora that reflect the region's ecological diversity.

1.11.6 Soil Types

The city's soils are primarily alluvial in origin, deposited by the fluvial activity of the Hooghly River over thousands of years. These soils are fertile, loamy, and well-drained, making them suitable for vegetable and fruit cultivation in peri-urban areas. The high organic content and fine texture of the alluvial deposits also contribute to the support of urban vegetation and landscape development projects.

1.11.7 Urban Development and Infrastructure

As one of India's oldest metropolitan centers, Kolkata exhibits a blend of colonial heritage and modern infrastructure. The Kolkata Municipal Corporation is responsible for overseeing essential services such as roads, bridges, public transportation, water supply, solid waste management, and sewage disposal. Despite rapid urbanization, challenges such as congestion, waste management inefficiencies, and aging infrastructure persist. The cityscape presents a striking contrast between colonial-era architecture, contemporary skyscrapers, and densely populated neighborhoods, reflecting its complex socio-spatial evolution.

1.11.8 Landmarks and Features

Kolkata is celebrated for its rich architectural and cultural heritage. Iconic landmarks include the Victoria Memorial, an elegant white marble monument symbolizing the city's colonial legacy, and the Howrah Bridge, a steel cantilever bridge that spans the Hooghly River and serves as a defining symbol of the city. Other notable sites include the Indian Museum, Marble Palace, St. Paul's Cathedral, and the ubiquitous yellow taxis, which have become cultural emblems of Kolkata's urban identity.

1.11.9 Transport and Connectivity

Kolkata enjoys robust transportation connectivity through an extensive network of roads, railways, and air routes. The Kolkata Metro, India's first underground rapid transit system, efficiently connects major parts of the city. In addition, the Netaji Subhas Chandra Bose International Airport provides both domestic and international air connectivity. The city's road network integrates national highways and state roads, while buses, trams, ferries, and suburban trains collectively form an accessible and affordable urban transport system.

1.11.10 Demography

Kolkata ranks among the most densely populated urban centers in India. According to the 2021 census estimates, the greater Kolkata region hosts nearly 14 million inhabitants, with approximately 4.5 million residents living within the KMC jurisdiction. The population is characterized by linguistic and cultural diversity, with Bengali serving as the predominant language. However, Hindi, English, Urdu, and several regional languages are also widely spoken. The city's demographic profile reflects its historical role as a center of migration, multicultural exchange, and intellectual activity.

In conclusion, the Kolkata Municipal Corporation area represents a vibrant and dynamic urban landscape defined by its historical heritage, demographic diversity, and environmental significance. Situated on the eastern bank of the Hooghly River, the city exhibits a unique interplay of natural and built environments, supported by its fertile alluvial plains and tropical climate. As one of India's principal economic, cultural, and administrative hubs, Kolkata continues to evolve while maintaining its distinctive character as the gateway to Eastern India.

Chapter 2
Spatio-Temporal Evaluation of Land Use and Green Space Changes

2.1 Background

The global human population is increasing continuously, while the availability of land for supporting both floral and faunal communities remains finite. This growing population exerts immense pressure on the earth's surface to meet essential needs such as food, housing, and infrastructure. Consequently, human settlements expand by encroaching upon the natural habitats of other species, leading to habitat loss, biodiversity decline, and ecological imbalance. These processes contribute significantly to changes in land use and land cover (LULC) patterns, triggering environmental challenges that threaten the long-term sustainability of human life. Although LULC transformation is a global phenomenon, its pace is exceptionally high in major urban centers due to rapid and often unregulated urbanization, particularly in megacities such as Mumbai, Delhi, Kolkata, Chennai, and Bengaluru. Urbanization, defined as the migration and concentration of people from rural to urban areas (Jat et al., 2008; Nourqolipour et al., 2016; Velmurugan & Sajjad, 2009), involves continuous spatial expansion and structural transformation of cities. It represents a dynamic and complex process that alters both the natural and built environments, resulting in the replacement of natural landscapes with impervious urban forms (Castle & Crooks, 2006; Dahal et al., 2018).

In developing countries, urban growth often occurs without adequate infrastructure or planning, leading to disorganized changes in land use and land cover (Rahaman et al., 2019). Such uncontrolled expansion produces unsystematic and unplanned urban development (Cohen, 2006; Grimm et al., 2008). The increasing demands of a growing population further accelerate these transformations, with land use evolving as a combined outcome of human actions and environmental conditions, while land cover reflects both natural processes and anthropogenic alterations (Reveshty, 2011; Mukhopadhaya, 2016). Urban identity and landscape configuration are shaped by diverse geographical, environmental, and socio-political drivers (Dutta et al., 2017; Sahana et al., 2018). In developing nations, metropolitan cities are expanding rapidly

© The Author(s), under exclusive license to Springer Nature Switzerland AG 2026
M. B. Ali and S. Jamal, *Urban Green Spaces and Environmental Sustainability*,
https://doi.org/10.1007/978-3-032-15322-7_2

due to population pressure and economic diversification (Al-Sharif & Pradhan, 2015; Luo & Mountrakis, 2010). Initially, urban expansion primarily affects peri-urban fringes, where LULC changes are most prominent (Nuissl et al., 2009). Over time, this process disrupts landscape functions and causes environmental degradation (Dewan & Yamaguchi, 2009; Jamil et al., 2018). Managing these transitions poses one of the greatest challenges for sustainable urban development in the twenty-first century (Cohen, 2004; Sakieh et al., 2017).

Technological advancements in remote sensing (RS) and Geographic Information Systems (GIS) have introduced sophisticated analytical techniques for studying urban growth dynamics (Awasthi et al., 2011; Al-Shalabi et al., 2013). These tools enable accurate mapping and temporal analysis of land cover changes (Haack & Rafter, 2006; Sahana et al., 2018) and support various models to interpret spatial patterns of LULC transformation (Al-Shalabi et al., 2013; Hashem & Balakrishnan, 2015; Jat et al., 2008; Prenzel, 2004).

Over the past five decades, India's total population has more than doubled, while its urban population has increased nearly fivefold (Rahaman et al., 2019). The last three decades have witnessed accelerated urbanization, particularly in Mumbai, Delhi, and Kolkata cities now ranked among the largest globally (Bhagat & Mohanty, 2009; Taubenbock et al., 2009). Although several studies (Mandal et al., 2019; Rahaman et al., 2019; Sahana et al., 2018) have examined urban growth and land transformation at the level of Kolkata's megacity and surrounding urban agglomerations, limited attention has been given to analyzing spatio-temporal LULC dynamics specifically within the Kolkata Municipal Corporation (KMC). The KMC area has undergone continuous spatial expansion over recent decades (Bhatta, 2009), warranting focused investigation.

Considering these aspects, the present chapter aims to evaluate the changing patterns of land use and land cover in KMC over a 30-year period (1990–2020). It also seeks to assess rapid built-up expansion, its underlying causes, and its relationship with relatively slower population growth. Using multi-temporal satellite datasets at 10-year intervals, the chapter endeavors to understand the dynamics of LULC transformation and urban growth in one of India's most complex and ecologically stressed metropolitan environments.

2.2 Database and Methodology

2.2.1 Data Sources

2.2.1.1 Spatial and Non-spatial Data

The present chapter utilizes both spatial and non-spatial datasets to analyze land use and land cover dynamics within the Kolkata Municipal Corporation (KMC) area. The spatial data comprise remote sensing imageries and the base map of KMC, while

Table 2.1 Non-spatial data used for this study

Year	Population	Growth (%)
1951	2,698,494	0.00
1971	3,148,746	16.70
1991	4,399,819	39.73
2011	4,496,694	2.20
**	6,000,000	

**Floating population per day
Source Census of India

non-spatial data have been extracted from the Census of India reports (Tables 2.1 and 2.2). To ensure temporal consistency and comparability, Landsat-5 Thematic Mapper (TM) satellite imageries for the years 1990, 2000, and 2010, along with Landsat-8 Operational Land Imager (OLI) imagery for 2020, have been employed in this analysis.

All satellite images correspond to the month of November, selected deliberately to minimize atmospheric disturbances such as cloud cover, haze, fog, and rainfall, which are typically minimal during this period. This choice ensures optimal visibility and radiometric clarity across all datasets. The imageries were obtained from the USGS Earth Explorer open-access platform (https://earthexplorer.usgs.gov). The specific path and row numbers used for all the selected imageries are 138 and 44, respectively.

Each satellite image possesses a spatial resolution of 30 m, allowing for detailed and accurate mapping of land cover features across the study area. The imagery was acquired at 10-year intervals, providing a consistent temporal framework for assessing patterns of land use/land cover transformation and built-up expansion within the KMC boundary. This uniform temporal and spatial configuration facilitates reliable comparison and analysis of changes over the three-decade period, ensuring precision in detecting the spatial dynamics and urban growth trends represented in this chapter.

For this study, cloud- and haze-free satellite imageries were carefully selected to ensure optimal image quality and reliable data interpretation. During the preprocessing stage, all images were georeferenced and geometrically corrected using the Universal Transverse Mercator (UTM) Zone 45N projection system, based on the World Geodetic System 1984 (WGS 84) datum. To achieve precise spatial alignment, image registration was performed employing the nearest neighbor resampling technique, which preserves the original spectral integrity of the data. A total of 30 Ground Control Points (GCPs) were utilized to ensure maximum overlay accuracy and spatial consistency across all temporal datasets (Ganaie et al., 2020).

2.2.1.2 Field Survey

To obtain ground-truth data for the validation of land use/land cover (LULC) classification, a GPS-based field survey was conducted across the study area. A total

Table 2.2 Details of spatial data used for this study

Year	Satellite	Sensor	Path/row	Spatial resolution (m)	Date	Wavelength (μm)
1990	Landsat-5	TM	138/44	30	14-Nov-1990	Band 1: 0.45–0.52
						Band 2: 0.52–0.62
						Band 3: 0.63–0.69
						Band 4: 0.76–0.90
						Band 5: 1.55–1.75
						Band 7: 2.08–2.35
2000	Landsat-5	TM	138/44	30	9-Nov-2000	Band 1: 0.45–0.52
						Band 2: 0.52–0.62
						Band 3: 0.63–0.69
						Band 4: 0.76–0.90
						Band 5: 1.55–1.75
						Band 7: 2.08–2.35
2010	Landsat-5	TM	138/44	30	5-Nov-2010	Band 1: 0.45–0.52
						Band 2: 0.52–0.62
						Band 3: 0.63–0.69
						Band 4: 0.76–0.90
						Band 5: 1.55–1.75
						Band 7: 2.08–2.35
2020	Landsat-8	OLI	138/44	30	16-Nov-2020	Band 1: 0.435–0.451
						Band 2: 0.452–0.511
						Band 3: 0.533–0.590
						Band 4: 0.636–0.673
						Band 5: 0.851–0.879
						Band 6: 1.566–1.651
						Band 7: 2.107–2.294

Source https://earthexplorer.usgs.gov

of 127 GPS coordinates were collected to verify and validate the classified imagery results. In addition, 108 ground reference points were extracted from Google Earth to further enhance the accuracy assessment and ensure spatial consistency between the classified outputs and actual ground conditions.

2.2.2 Methods

The systematic methodological framework adopted for this research is outlined in the flowchart (Fig. 2.1), which illustrates the sequential steps followed during data processing, classification, and validation.

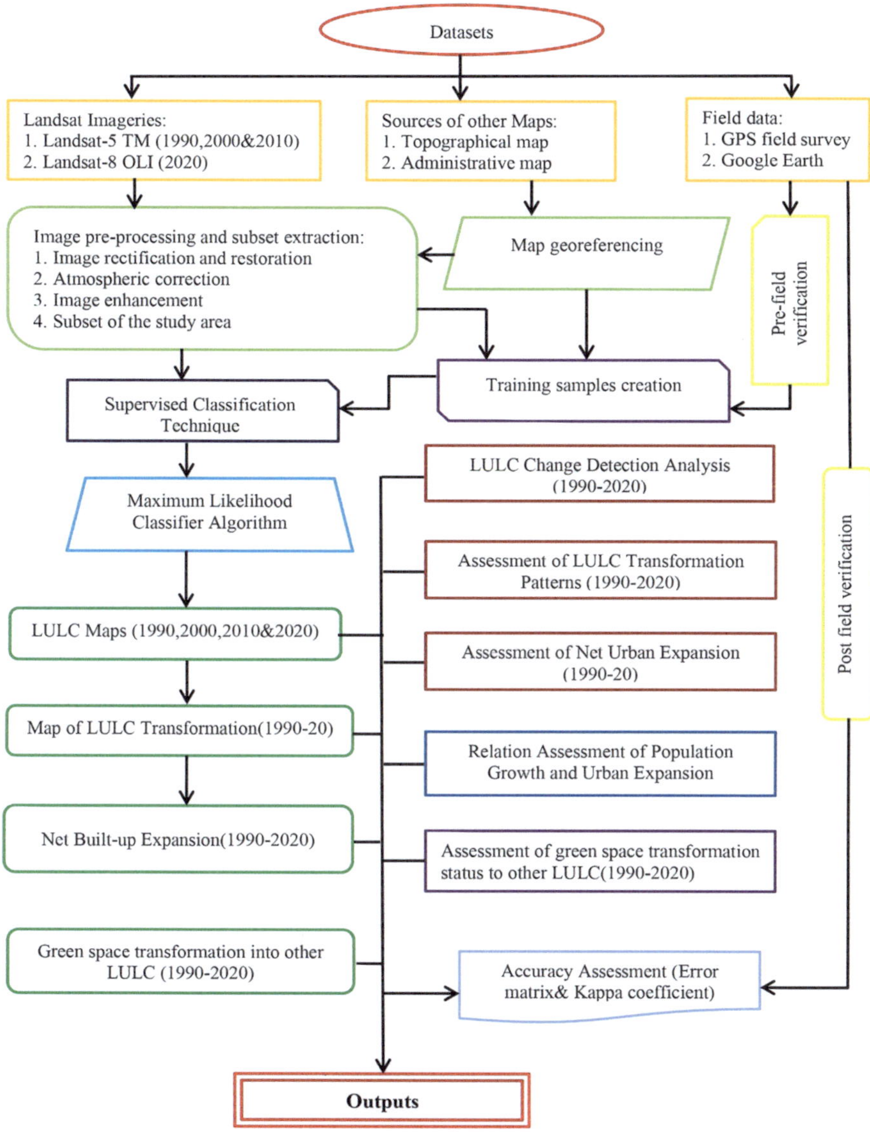

Fig. 2.1 Methodological flowchart of the chapter

Table 2.3 Land use/land cover classification scheme and description of classes

LULC types	Description
Built-up area	Residential, commercial and services, industrial, transportation, roads, and other man-made structures
Water body	River, permanent open water, lakes, ponds, reservoirs, permanent and seasonal wetlands, marshy land, and swamps
Green space	Deciduous forest, mixed forest, grassland, parks, community gardens, scrub, commercial horticulture plantations, orchards, rooftop plants, and others vegetative covers
Fallow land	Exposed soils, bare land, landfill sites, areas of active excavation, sand and transitional area between water body and mainland with never have more than 10% vegetated cover during any time of the year

Source Dewan & Yamaguchi (2009), Roy et al. (2016)

2.2.2.1 Supervised Image Classification Method

Initially, a subset of the Area of Interest (AOI) was extracted from each satellite image tile for detailed analysis. The AOI was then subjected to supervised classification using 100 training samples per class, carefully selected and cross-verified with Google Earth imagery and Survey of India topographical sheets. The land use/land cover categories considered in this study include built-up area, water body, fallow land, and green space (Table 2.3).

For the classification process, the Maximum Likelihood algorithm, a statistically robust and widely used method, was applied to differentiate between the identified classes. Using this approach, four temporal datasets corresponding to 1990, 2000, 2010, and 2020 were classified, each representing distinct stages of urban expansion and land cover transformation. The classified LULC maps were then generated using ERDAS IMAGINE 2014 and ArcMap 10.5, which provided reliable spatial and thematic outputs for subsequent analysis.

2.2.2.2 Land Use/Land Cover Change Detection Method

To analyze the land use/land cover (LULC) transformation across three decades, the quantitative values of each land use category were employed to calculate the rate of change in respective LULC classes. The computation was carried out using the following formula as suggested by Sahana and Sajjad (2019):

$$R = \left[\frac{1}{t_1 - t_2}\right] \times \left[\text{ln}\left(\frac{C_1}{C_2}\right)\right],$$

where R represents the rate of land use/land cover change; C_1 and C_2 represent area under LULC categories in time periods t_1 (2020) and t_2 (1990) respectively.

2.2.2.3 Calculation of Built-Up Expansion Rate

The built-up expansion rate (BER) for the study area was computed to determine the pace of urban growth over the three temporal intervals 1990–2000, 2000–2010, and 2010–2020. The following equation was used:

$$\mathrm{BER} = \frac{B_aT_2 - B_aT_1}{B_aT_1} \times 100\%,$$

where BER = urban expansion rate, B_aT_1 = pixel counts of urban area map first, B_aT_2 = pixel counts of urban area map second.

2.2.2.4 Calculation of Net Built-Up Expansion Rate

In addition to the expansion rate, the net built-up expansion rate (NBAE) was calculated to quantify the total areal increase in urbanized land between two time intervals. The computation follows the equation given below:

$$\mathrm{NBAE} = \frac{B_aT_2 - B_aT_1}{B_aT_1} \times 100\%,$$

where NBAE = net built-up area expansion, B_aT_1 = pixel counts of built-up area first, B_aT_2 = pixel counts of built-up area second.

2.2.2.5 Linear Regression

To establish the relationship between population growth and built-up expansion, a scatter plot with a linear regression line was constructed following Mandal et al. (2019). The analysis was performed using MS Excel 2010, applying the following regression equation:

$$Yc = a + b,$$

where Yc represents estimated value, a = y-intercept, b = regression coefficient.

The constants a and b are determined by:

$$b = \frac{N\sum XY - \sum X \sum Y}{N\sum X^2 - \left(\sum X\right)^2}, \quad a = \frac{\sum Y - b\sum X}{N}.$$

2.2.2.6 Calculation of the Decadal Population

The population projection for the years 1990, 2000, 2010, and 2020 was calculated using Census of India data to facilitate the analysis of urban growth and built-up expansion across the three decades. The following equation was employed to project the population for each respective period:

$$P_{pj} = P_2 + y\frac{(P_1 - P_2)}{N},$$

where P_{pj} = population projection, P_1 = present total population according to the recent census, P_2 = total population in the previous census, y = number of years from the previous census, and N = total number of years between the recent and previous census.

2.2.2.7 Assessment Accuracy and Validation of Maps

Accuracy assessment is a crucial step in verifying the reliability of classified images by comparing them with actual ground-truth data. The Kappa coefficient (κ), a discrete multivariate statistical measure, was employed to evaluate the accuracy of the LULC classifications for the years 1990, 2000, 2010, and 2020 (Cohen, 1960; Kar et al., 2018). The Cohen's Kappa coefficient was computed using the following expression:

$$\kappa = \frac{n\sum x_{ii} - \sum x_{i+}x_{+i}}{n^2 - \sum x_{i+}x_{+i}},$$

where r = number of rows in the matrix, x_{ii} = number of observations in row i and column i, x_i+ and x_{+i} = marginal totals of row i and column i, respectively, and n = total number of observations (samples/pixels).

This statistical validation ensured that the classified land use/land cover maps produced for each time period were accurate and reliable for subsequent spatial and temporal analyses.

2.3 Results and Discussion

2.3.1 Historical Land Use and Land Cover (LULC) Transformations in Kolkata Over the Last Three Decades (1990–2020)

2.3.1.1 Report of Accuracy and Validation

The accuracy of the land use/land cover (LULC) classification was evaluated using ERDAS IMAGINE 2014, employing Ground Control Points (GCPs) derived from Google Earth, Survey of India Topographical sheets, and GPS-based field observations. The Kappa statistic, a widely recognized measure of classification reliability, was used to assess the degree of agreement between the classified and reference data. According to Lillesand et al. (2015), the Kappa coefficient ranges between 0 and 1, where a value of 0 indicates agreement due to chance, 0.01–0.20 indicates slight agreement, 0.21–0.40 fair agreement, 0.41–0.60 moderate agreement, 0.61–0.80 substantial agreement, 0.81–0.99 near-perfect agreement, and 1.00 represents perfect agreement (Mandal et al., 2019).

For this study, four temporal satellite imageries corresponding to the years 1990, 2000, 2010, and 2020 were classified into four major categories: built-up area, water bodies, fallow land, and green spaces. Each classified image underwent accuracy evaluation using 200 training sample points derived from Google Earth GCPs. Due to data availability constraints, Google Earth was the only feasible source for validating the historical imagery of 1990, 2000, and 2010, whereas the 2020 classification was verified through both GPS-based ground-truth data and Google Earth control points.

The accuracy assessment confirmed that the classifications achieved substantial to near-perfect agreement, as reflected in the calculated Kappa coefficient and overall accuracy values. Specifically, the Kappa values were 0.74 for 1990, 0.78 for 2000, 0.77 for 2010, and 0.84 for 2020, indicating a consistently high level of classification reliability. Correspondingly, the overall classification accuracies were 81.46% (1990), 83.96% (2000), 82.98% (2010), and 91.64% (2020), respectively (Table 2.4). These results validate the robustness of the classification process and confirm that the LULC maps produced for this study are reliable representations of the spatial and temporal land cover patterns across the Kolkata Municipal Corporation.

2.3.1.2 Dynamics of Land Use/Land Cover Patterns in KMC from 1990 to 2020

Change is an inherent and inevitable process in nature, and over time, every spatial entity experiences transformation. The Kolkata Municipal Corporation (KMC) area is no exception to this natural phenomenon. Over the past three decades, the city has

Table 2.4 Accuracy assessment statistics of maps

	Category	Year			
		1990	2000	2010	2020
User accuracy (%)	Built-up area	83.5	81.2	80.1	91.4
	Water body	85.3	91.4	89.9	93.9
	Fallow land	72.4	79.7	74.6	85.5
	Green space	81.9	82.3	82.4	91.4
Producer accuracy (%)	Built-up area	85.3	78.5	83.3	93.2
	Water body	86.1	90.1	91.5	94.6
	Fallow land	71.3	81.2	79.2	88.2
	Green space	85.9	87.3	82.8	94.9
Overall Kappa statistics		0.74	0.78	0.77	0.84
Overall classification accuracy (%)		81.46	83.96	82.98	91.64

undergone significant modifications in its land use and land cover (LULC) configuration, reflecting the intensity of urbanization and associated environmental shifts (Fig. 2.2). A comparative assessment of the LULC maps for 1990 and 2020 reveals a striking contrast in the spatial distribution of land cover categories within the KMC boundary (Fig. 2.2). These temporal analyses indicate that all major LULC classes namely built-up area, water bodies, green spaces, and fallow land have experienced substantial changes, with the most pronounced transformation occurring in the built-up category. During the 30-year period from 1990 to 2020, the built-up area expanded significantly from 9094.90 to 13,199.29 ha, illustrating the accelerated pace of urban growth. In contrast, water bodies declined from 2526.71 to 1448.35 ha, fallow land decreased from 1363.50 to 675.48 ha, and green spaces contracted from 7999.25 to 5661.23 ha. These figures demonstrate a consistent pattern of urban encroachment and the gradual loss of ecologically vital areas within the municipal limits (Table 2.5). Overall, the analysis of land use/land cover dynamics over the three decades highlights a clear trajectory of urban expansion at the expense of natural and semi-natural landscapes. This trend underscores the urgent need for environmentally conscious urban planning and the conservation of remaining green and open spaces to sustain ecological balance and urban resilience in the Kolkata Municipal Corporation area.

A careful examination of Table 2.6 reveals a pronounced transformation in the land use and land cover (LULC) patterns during the initial decade of the study period, extending from 1990 to 2000. During this interval, the built-up area expanded markedly from 9094.90 ha in 1990 to 10,268.53 ha in 2000 representing an absolute increase of 1173.63 ha or a 12.90% growth in urbanized land. In contrast, other LULC categories such as water bodies, fallow land, and green spaces exhibited noticeable reductions. The area under water bodies declined from 2526.71 to 2324.77 ha, fallow land decreased from 1363.50 to 1259.49 ha, and green spaces diminished from 7999.25 to 7131.56 ha. Quantitatively, this equates to a loss of 201.94 ha (7.99%) of water bodies, 104.01 ha (7.63%) of fallow land, and 867.69 ha (10.85%) of green

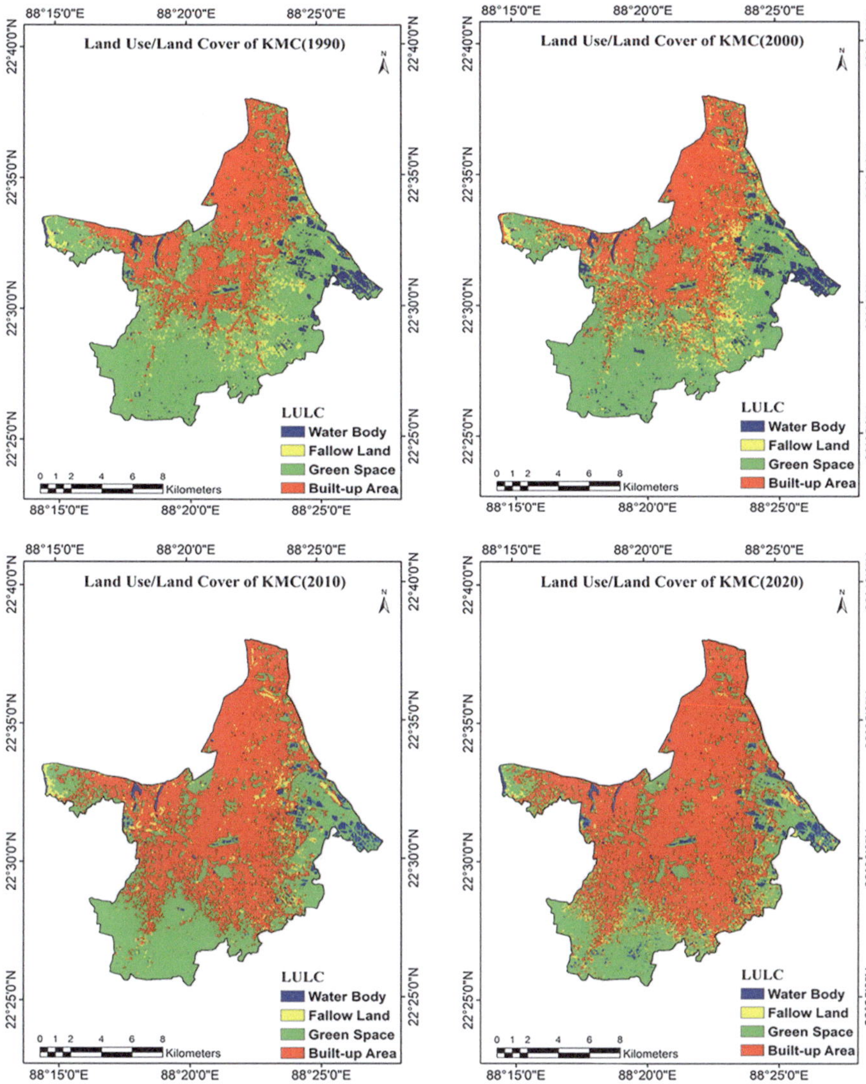

Fig. 2.2 Spatio-temporal land use/land cover pattern of KMC for 1990, 2000, 2010, and 2020

spaces over the decade. The LULC maps of 1990 and 2020 (Fig. 2.2) further illustrate the spatial manifestation of these changes, showing a clear expansion of the built-up area into adjacent ecological zones. Urban development has progressively encroached upon green spaces, wetlands, and vacant lands, with the southern-central

Table 2.5 Spatio-temporal land use/land cover statistics (1990–2020)

LULC category	1990		2000		2010		2020		Change (1990–2020)	Change
	ha	%	ha	%	ha	%	ha	%	ha	(%)
Built-up area	9094.90	43.34	10,268.53	48.93	12,337.71	58.79	13,199.29	62.90	4104.39	45.13
Water body	2526.71	12.04	2324.77	11.08	1838.52	8.76	1448.35	6.90	−1078.36	−42.68
Fallow land	1363.50	6.50	1259.49	6.00	864.28	4.12	675.48	3.22	−688.02	−50.46
Green space	7999.25	38.12	7131.56	33.99	5943.85	28.33	5661.23	26.98	−2338.02	−29.23
Total	20,984.35	100.00	20,984.35	100.00	20,984.36	100.00	20,984.35	100.00		

sector of the Kolkata Municipal Corporation particularly affected by the conversion of vegetated areas into built-up surfaces. This pattern underscores the accelerating urban sprawl and the consequent depletion of environmentally sensitive land categories within the city.

The decade spanning 2000–2010 marked a period of pronounced transformation in the land use and land cover (LULC) configuration of the Kolkata Municipal Corporation (KMC). During this interval, there was a substantial contraction in the extent of water bodies, fallow land, and green spaces, reflecting the intensifying pace of urban expansion. Specifically, the area under water bodies declined from 2324.77 to 1838.52 ha, fallow land decreased from 1259.49 to 864.28 ha, and green spaces reduced from 7131.56 to 5943.85 ha. These changes correspond to absolute losses of 486.25, 395.21, and 1187.71 ha, respectively, representing percentage decreases of −20.92, −31.38, and −16.65% (Table 2.7). In contrast, the built-up area exhibited remarkable growth, increasing from 10,268.53 ha in 2000 to 12,337.71 ha in 2010, an absolute rise of 2069.18 ha, equivalent to a 20.15% expansion. The LULC maps for 2000 and 2010 (Fig. 2.2) vividly illustrate this trend, indicating that new built-up developments predominantly extended toward the southern, southeastern, and southwestern zones of the KMC area. This spatial pattern of urban growth underscores the ongoing process of urban sprawl and land conversion, where ecologically important spaces such as water bodies and vegetation are increasingly being replaced by impervious built-up surfaces, thereby intensifying environmental stress within the city.

Table 2.6 Change in the area of different LULC classes (1990–2000)

LULC category	1990		2000		Changes (1990–2000)	
	ha	%	ha	%	ha	%
Built-up area	9094.90	43.34	10,268.53	48.93	1173.63	12.90
Water body	2526.71	12.04	2324.77	11.08	−201.94	−7.99
Fallow land	1363.50	6.50	1259.49	6.00	−104.01	−7.63
Green space	7999.25	38.12	7131.56	33.99	−867.69	−10.85
Total	20,984.35	100.00	20,984.35	100.00		

Table 2.7 Change in the area of different LULC classes (2000–2010)

LULC category	2000		2010		Changes (2000–2010)	
	ha	%	ha	%	ha	%
Built-up area	10,268.53	48.93	12,337.71	58.79	2069.18	20.15
Water body	2324.77	11.08	1838.52	8.76	−486.25	−20.92
Fallow land	1259.49	6.00	864.28	4.12	−395.21	−31.38
Green space	7131.56	33.99	5943.85	28.33	−1187.71	−16.65
Total	20,984.35	100.00	20,984.36	100.00		

Similarly, Fig. 2.2 depicts the transformation of land use and land cover (LULC) patterns during the decade from 2010 to 2020, highlighting the continued dominance of urban expansion across the Kolkata Municipal Corporation (KMC) area. Over this period, the built-up area exhibited sustained positive growth, increasing from 12,337.71 ha in 2010 to 13,199.29 ha in 2020 an overall gain of 861.58 ha, representing a 6.98% rise. In contrast, other LULC categories such as water bodies, fallow lands, and green spaces recorded further declines, reflecting the pressure of encroaching urban development. The extent of water bodies contracted from 1838.52 to 1448.35 ha, marking a loss of 390.17 ha (−21.22%), while fallow land diminished from 864.28 to 675.48 ha, a decrease of 188.80 ha (−21.84%). Similarly, green spaces were reduced from 5943.85 to 5661.23 ha, a decline of 282.62 ha (−4.75%) (Table 2.8). The continued reduction in these land categories can be attributed primarily to the encroachment and expansion of built-up areas, which have progressively consumed open and ecologically sensitive lands. The LULC map of 2010 clearly indicates that substantial tracts of fallow land, green cover, and water bodies were transformed into urban settlements by 2020 (Fig. 2.2). The 2020 LULC map further emphasizes the spatial concentration of urban sprawl, showing that the most extensive growth occurred toward the southeastern, southern, southwestern, and western sectors of the KMC. This continuing expansion underscores the persistent challenge of managing urban growth while conserving the city's diminishing natural and semi-natural environments.

The spatio-temporal analysis of land use and land cover (LULC) over the past three decades reveals a pronounced and continuous expansion of the built-up area within the Kolkata Municipal Corporation (KMC), reflecting the city's rapid pace of urbanization. Between 1990 and 2020, the total extent of built-up land increased by 4104.39 ha, amounting to a remarkable 45.13% rise in urban coverage (Figs. 2.3 and 2.4, Table 2.5). Conversely, the city experienced a steep decline in other major land cover categories. The total area under water bodies, fallow lands, and green spaces decreased by 1078.36 ha (−42.68%), 688.02 ha (−50.46%), and 2338.02 ha (−29.23%), respectively (Table 2.5). These figures underscore the growing dominance of built-up land over natural and semi-natural environments. Field observations further revealed that several locations once categorized as open or fallow land have been formally converted into public parks and gardens, many of which are now maintained jointly by municipal authorities and private organizations. However, numerous

Table 2.8 Change in the area of different LULC classes (2010–2020)

LULC category	2010		2020		Changes (2010–2020)	
	ha	%	ha	%	ha	%
Built-up area	12,337.71	58.79	13,199.29	62.90	861.58	6.98
Water body	1838.52	8.76	1448.35	6.90	−390.17	−21.22
Fallow land	864.28	4.12	675.48	3.22	−188.80	−21.84
Green space	5943.85	28.33	5661.23	26.98	−282.62	−4.75
Total	20,984.36	100.00	20,984.35	100.00		

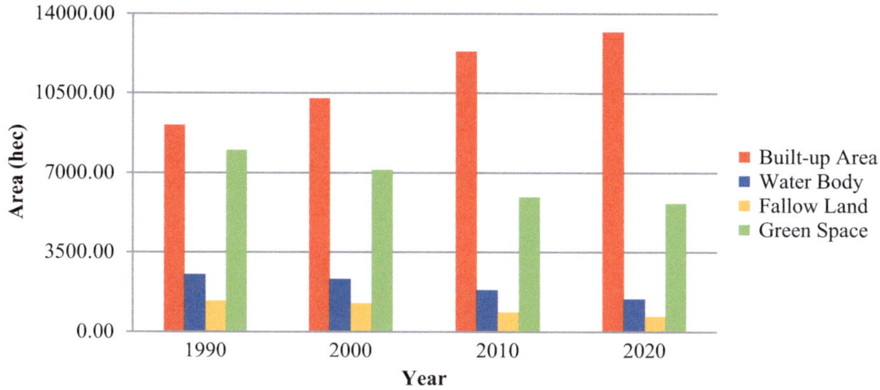

Fig. 2.3 Spatio-temporal land use/land cover pattern in KMC (1990–2020)

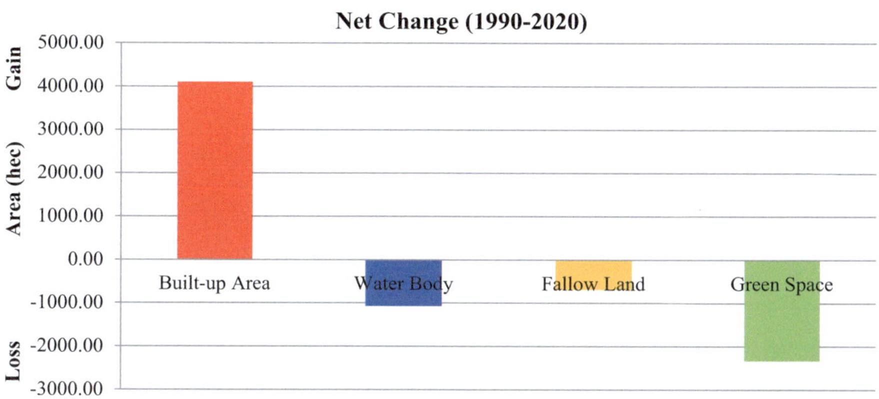

Fig. 2.4 Spatio-temporal land use/land cover classes gain and loss in KMC over last three decades

small water bodies, particularly in the southern, southeastern, western, and central zones of the KMC, have been lost to urban development, with some converted into built-up areas or landscaped green patches. The degradation and transformation of these natural water bodies are closely associated with urban encroachment, global warming, shifting climatic patterns, and a decline in mean annual rainfall, collectively contributing to the city's environmental stress and hydrological imbalance.

2.3.1.3 Trend Analysis of Land Use/Land Cover from 1990 to 2020

This section presents the trend analysis of various land use and land cover (LULC) categories for the Kolkata Municipal Corporation (KMC) area from 1990 to 2020, employing regression analysis to determine the strength and direction of relationships through coefficient values. Over the span of three decades, each LULC class displays a distinct trajectory, as illustrated in Fig. 2.5. The scatter plots reveal that water bodies, fallow lands, and green spaces exhibit a declining trend, with strong negative correlations and coefficient of determination (R^2) values of 0.97, 0.95, and 0.95, respectively. These results highlight a consistent reduction in the area of these categories, indicating the ongoing loss of natural and semi-natural spaces within the urban landscape. In contrast, the built-up area demonstrates a strong positive correlation, with an R^2 value of 0.97, reflecting a steady and substantial expansion throughout the study period (Fig. 2.5). The high coefficient values across all categories suggest that the transformations are statistically significant and strongly associated with rapid urbanization. The observed decline in water bodies, fallow lands, and green spaces corresponds to the concurrent increase in built-up areas, implying that the latter has expanded largely at the expense of the former. These changes are primarily driven by anthropogenic pressures, including intensifying human activities, urban infrastructure development, and environmental factors such as climate change and global warming, all of which have collectively reshaped the city's ecological and physical environment over the past three decades.

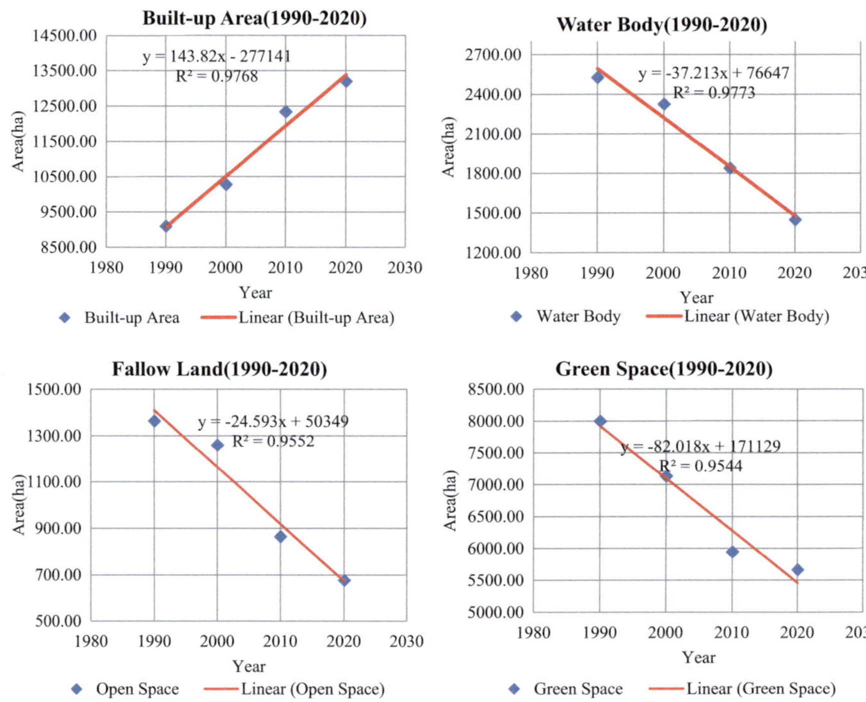

Fig. 2.5 Trend analysis of land use/land cover in KMC from 1990 to 2020

2.3.1.4 Land Use/Land Cover Transformation Patterns (1990–2020)

The land use and land cover (LULC) transformation map (Fig. 2.6) for the period 1990–2020 vividly portrays the spatial transitions that have occurred across various LULC categories within the Kolkata Municipal Corporation (KMC) over the past three decades. To quantify these changes, a transformation matrix was developed, summarizing the shifts between different land cover types during this timeframe. Table 2.9 provides a detailed account of both changed and unchanged areas, offering insight into the magnitude and direction of these transformations. The analysis reveals that ~361.27 ha of land remained unchanged within the fallow land category, while 245.45, 692.12, and 64.66 ha of fallow land were converted into built-up areas, green spaces, and water bodies, respectively (Table 2.9). Similarly, an estimated 4119.83 ha of land persisted as green space throughout the three decades. However, significant portions of this green area around 286.62, 3515.71, and 77.09 ha were transformed into fallow land, built-up zones, and water bodies, respectively (Table 2.9). In the case of water bodies, ~1306.61 ha remained unchanged during the study period, while 27.59, 343.23, and 849.28 ha transitioned into fallow land, built-up land, and green spaces, respectively (Table 2.9). These findings collectively underscore the extensive and persistent conversion of natural and semi-natural landscapes—particularly green spaces and water bodies into built-up areas, reflecting the continuous pressure of urban expansion and anthropogenic transformation in Kolkata over the last 30 years.

2.3.2 Geospatial Analysis of Urban Expansion and Its Impact on Green Space Depletion

2.3.2.1 Transformation of Green Space into Land Use Land Covers (1990–2020)

The transformation of green spaces into other land use and land cover (LULC) categories between 1990 and 2020 provides a clear depiction of the extent and direction of landscape change within the Kolkata Municipal Corporation (KMC). As presented in Figs. 2.7 and 2.8 and summarized in Table 2.10, ~3515.71 ha of green space have been converted into built-up areas, while 286.62 ha have transitioned into fallow lands, and another 77.09 ha have been transformed into water bodies. Among these transitions, the conversion of green spaces into built-up areas represents the most substantial and concerning trend, reflecting the pressure exerted by rapid urbanization and population growth. This process is primarily driven by the increasing demand for residential housing, infrastructure, and the establishment of economic and administrative facilities, including commercial centers, healthcare institutions, and transportation networks, to support the expanding urban population. The second-largest transformation, involving the conversion of green spaces into fallow lands, suggests a stage of transitional land use, where currently idle or semi-utilized lands

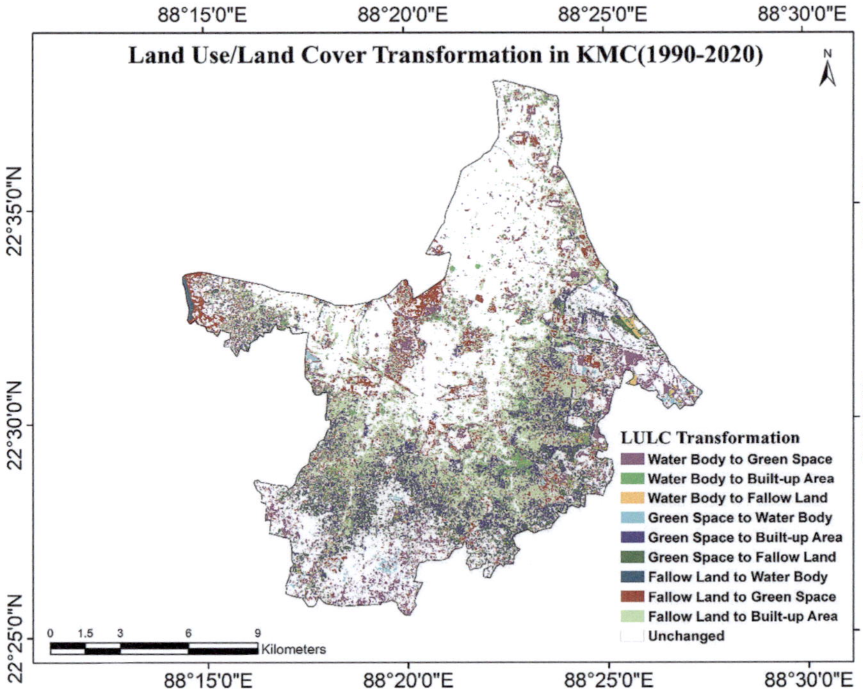

Fig. 2.6 Land use/land cover transformation in KMC from 1990 to 2020

Table 2.9 Land use/land cover transformation matrix (1990–2020)

LULC classes		LULC classes 2020				Total
		Fallow land	Built-up area	Green space	Water body	1990
LULC classes 1990	Fallow land	**361.27**	245.45	692.12	64.66	1363.50
	Built-up area	0.00	**9094.90**	0.00	0.00	9094.90
	Green space	286.62	3515.71	**4119.83**	77.09	7999.25
	Water body	27.59	343.23	849.28	**1306.61**	2526.71
	Total (2020)	675.48	13,199.29	5661.23	1448.35	20,984.35

Bold values represent unchanged area

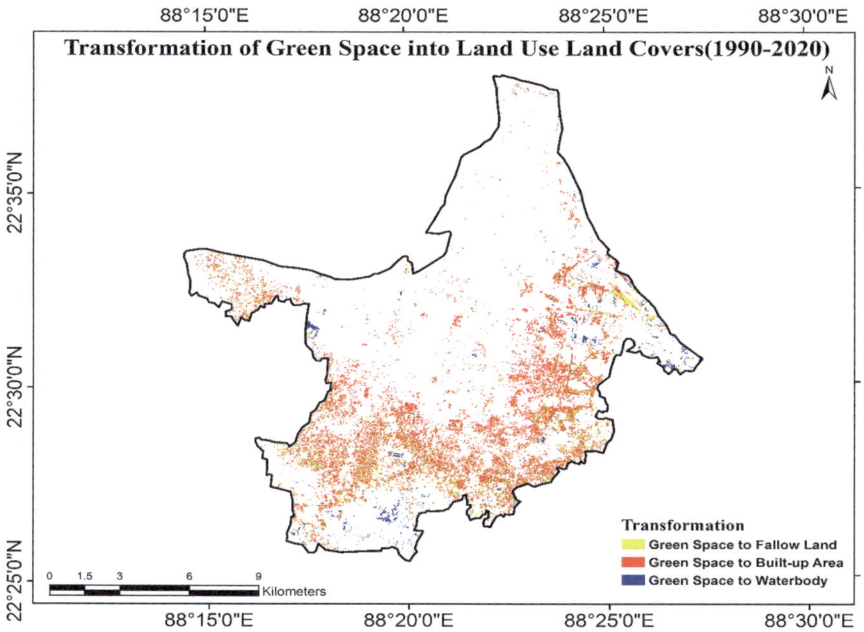

Fig. 2.7 Land use/land covers transformation into built-up (1990–2020)

may, in the near future, be further developed into urban built-up zones. Such patterns highlight the ongoing urban sprawl and the progressive loss of ecologically significant green spaces, emphasizing the urgent need for sustainable land management strategies and green infrastructure planning to balance development with environmental conservation within the city.

2.3.2.2 Spatio-Temporal Patterns of Built-Up Expansion

Figure 2.9 vividly illustrates the spatio-temporal dynamics of built-up area expansion in the Kolkata Municipal Corporation (KMC) between 1990 and 2020, highlighting the rapid urban growth patterns within the city. A close examination of the figure reveals that the southern, southeastern, and southwestern zones of the KMC have undergone the most pronounced and continuous expansion of built-up areas during this period. Conversely, the northern, northeastern, and northwestern parts were already extensively urbanized by 1990, and therefore exhibited relatively minimal change thereafter (Fig. 2.9). The eastern segment of the city, however, has remained largely undeveloped and ecologically preserved, owing to the presence of the East Kolkata Wetlands, a designated Ramsar Site that is legally protected and managed under the provisions of the East Kolkata Wetlands (Conservation and Management) Act, 2006. On the western boundary, urban growth has been comparatively restricted

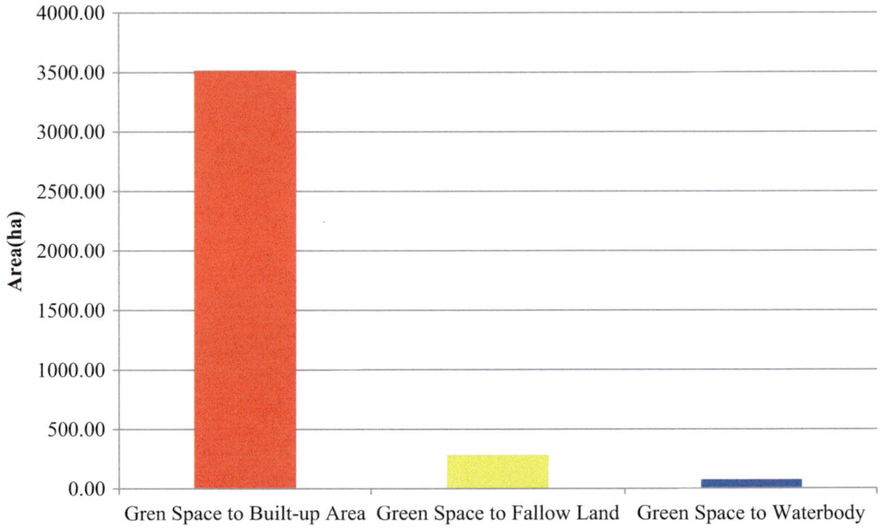

Fig. 2.8 Transformation of green space into different land use/land covers (1990–2020)

Table 2.10 Transformation of green space into different land use/land covers (1990–2020)

Transformation	Area (1990–2020)
Green space to built-up area	3515.71
Green space to fallow land	286.62
Green space to water body	77.09

due to the natural barrier posed by the Hooghly River, which limits spatial expansion in that direction. Similarly, the northwestern section of the city encompasses a prominent green belt, consisting of significant open spaces such as the Maidan, Race Course, Victoria Memorial, and Alipore Zoological Garden, all of which have remained free from built-up encroachment (Fig. 2.9). In the central region, several open spaces and recreational areas, including the Tollygunge Golf Club, Golf Garden, and Dhakuria Lake, have also persisted as non-built-up zones, serving as vital ecological and recreational buffers within the densely urbanized cityscape. Collectively, these areas represent the few remaining unchanged and preserved landscapes within the highly urbanized core of Kolkata, maintained primarily due to their ecological, recreational, and legal significance.

2.3.2.3 Built-Up Expansion Rate

The rate of built-up area expansion within the Kolkata Municipal Corporation (KMC) over the 30-year period from 1990 to 2020 demonstrates a pronounced and accelerating trend of urban growth, as illustrated in Fig. 2.9 and summarized in Table 2.11.

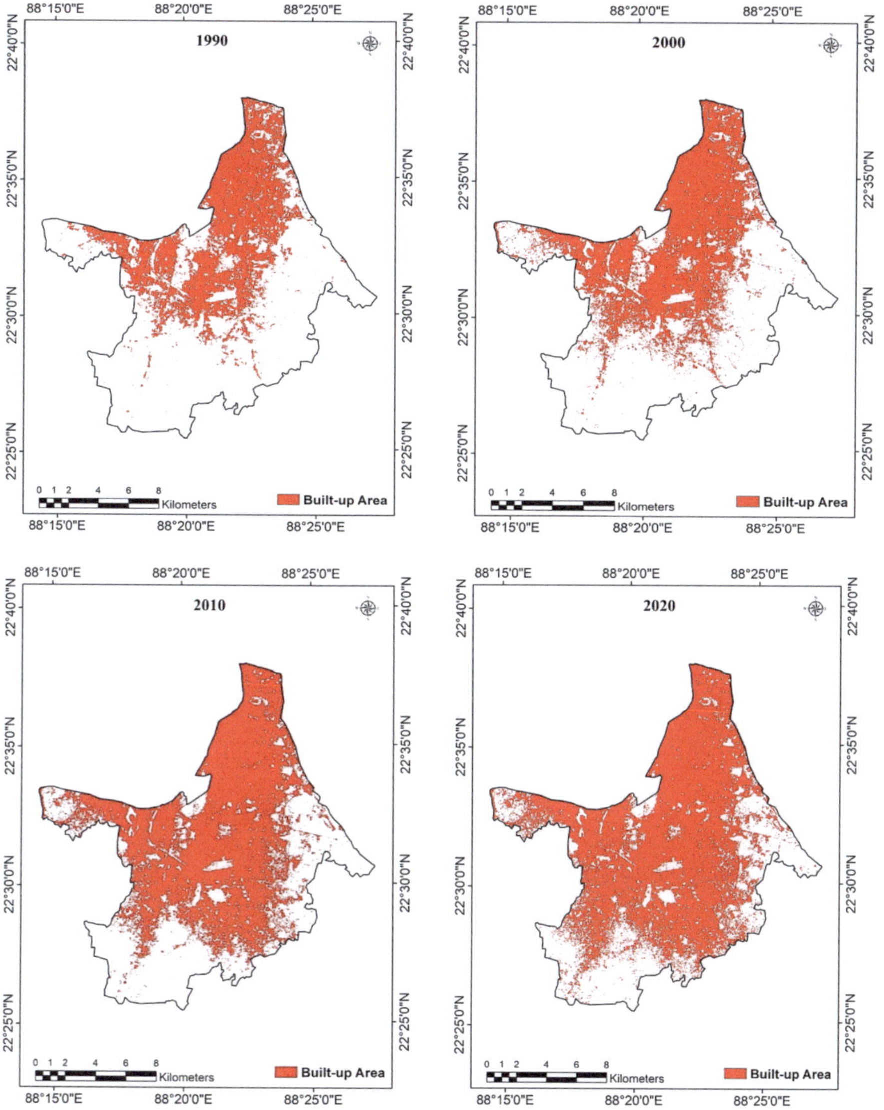

Fig. 2.9 Built-up expansion in KMC between the years 1990 and 2020

Although the expansion remained consistently high throughout the study period, the rate of increase varied notably across the three decades. During the first decade (1990–2000), the built-up area expanded from 9094.90 to 10,268.53 ha, marking an increase of 12.90%. This growth rate intensified during the second decade (2000–2010), when the built-up area surged from 10,268.53 to 12,337.71 ha, reflecting a substantial 20.15% rise the highest rate recorded across the study period. In the final decade (2010–2020), although the rate of expansion slightly decelerated compared

Table 2.11 Decadal population growth and urban expansion (1990–2020)

Year	Population	Built-up area/urban expansion
1990*	4,337,265	9094.90
2000*	4,443,413	10,268.53
2010*	4,491,850	12,337.71
2020*	4,540,288	13,199.29

*Calculated decadal population based on census data

to the previous period, it still remained significant at 6.98%, corresponding to an increase from 12,337.71 to 13,199.29 ha. When examined cumulatively, the overall net built-up expansion over the three decades amounts to an impressive 45.13%, translating to a total areal growth of 4104.39 ha within the KMC boundary (Fig. 2.10, Table 2.11). This consistent upward trajectory in built-up expansion underscores the rapid pace of urbanization and infrastructural development in Kolkata, driven by demographic pressures, economic growth, and land conversion from natural and semi-natural categories to built-up zones.

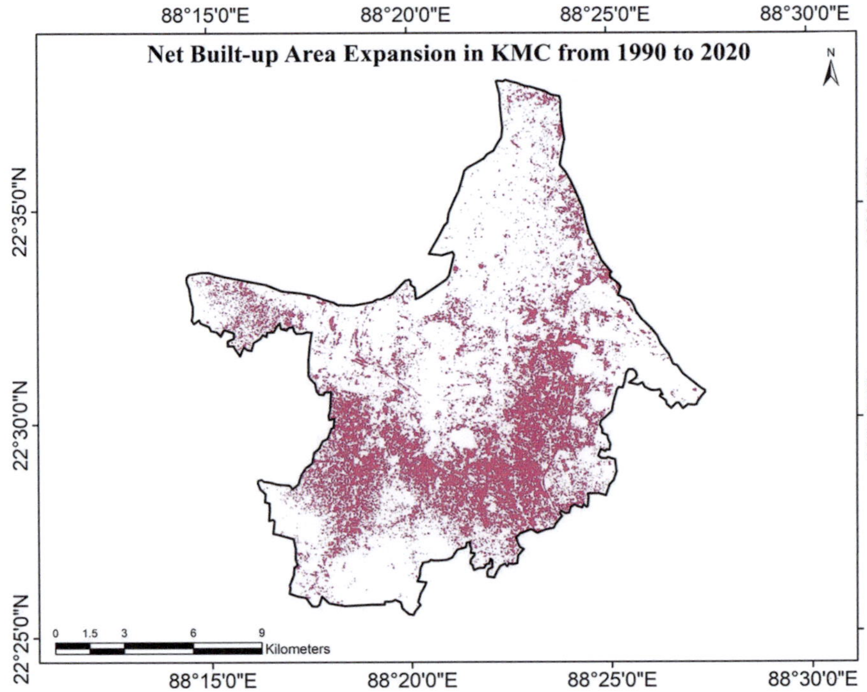

Fig. 2.10 Net built-up expansion in KMC from 1990 to 2020

2.3.2.4 Driving Forces Behind LULC Transformation and Built-Up Expansion

Table 2.11 clearly demonstrates the relationship between the rapid expansion of built-up areas and the steady population growth within the Kolkata Municipal Corporation (KMC) between 1990 and 2020. Over this 30-year span, the built-up area increased substantially from 9094.90 to 13,199.29 ha, reflecting an intense phase of urban expansion and infrastructural development. However, this growth in built-up land has not been proportionate to the rate of population increase during the same period (Table 2.11). As the third-largest megacity in India and the capital of West Bengal, Kolkata accommodates an immense permanent population of ~4.5 million, along with an additional floating population of around 6 million, which notably surpasses its residential population (Table 2.1). This floating population, consisting of daily commuters, informal workers, and transient residents, exerts significant pressure on urban land resources, accelerating the conversion of natural and semi-natural landscapes into built-up environments to meet the growing demand for housing, infrastructure, administrative facilities, commercial establishments, and employment hubs. Consequently, population dynamics both permanent and transient emerge as the dominant driving force behind the observed land use and land cover (LULC) transformation in the KMC. This extensive transformation has led to the depletion of vital environmental components, including green spaces, water bodies, and fallow lands, and has contributed to a host of urban environmental challenges such as biodiversity loss, air and noise pollution, soil degradation, microclimatic alterations, and the intensification of urban heat island effects. These patterns highlight the urgent need for sustainable urban planning and eco-sensitive governance to balance the city's developmental trajectory with ecological preservation.

2.3.2.5 Relationship Between Population Growth and Built-Up Expansion

To analyze the relationship between population growth and the expansion of built-up areas within the Kolkata Municipal Corporation (KMC), a regression analysis was conducted using scatter plots and a linear regression model. This statistical approach helped to quantify the degree of association between the two variables and to visually interpret the trend over the study period. The results of the regression analysis indicate a strong positive correlation between population growth and built-up area expansion, suggesting that increases in population directly influence urban spatial development patterns. The regression equation derived from the analysis shows a correlation coefficient of 0.92, while the coefficient of determination ($R^2 = 0.92$) confirms that ~92% of the variation in built-up area expansion can be explained by population growth (Fig. 2.11). This strong statistical relationship underscores the pivotal role of demographic pressure as a key driving force behind urban land transformation, reinforcing the need for integrated urban planning strategies that accommodate population growth without compromising environmental sustainability.

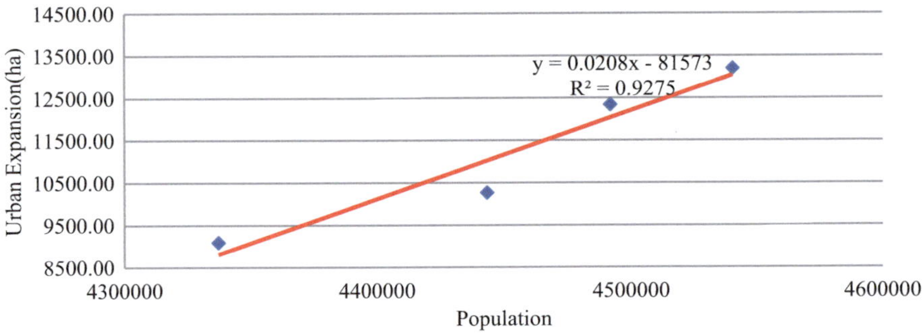

Fig. 2.11 Relationship between the growth of population and built-up expansion

This chapter presents a comprehensive quantitative analysis of the spatio-temporal dynamics of land use and land cover (LULC) within the Kolkata Municipal Corporation (KMC), which has experienced a significant expansion of built-up areas between 1990 and 2020 as a consequence of rapid population growth and economic development. Employing an integrated approach that combines geospatial techniques and statistical analyses, the study ensures the generation of accurate and empirically grounded results. Over the 30-year period, remarkable transformations in LULC have been observed, including a net increase in built-up areas by 4104.39 ha, accompanied by substantial declines in fallow lands (−688.02 ha), water bodies (−1078.36 ha), and green spaces (−2338.02 ha). The transformation matrix further highlights that the most substantial conversions occurred from green spaces to built-up areas (3515.71 ha), followed by water bodies to green spaces (849.28 ha) and fallow lands to green spaces (692.12 ha). The built-up area expanded from 9094.90 ha in 1990 to 13,199.29 ha in 2020, with the southern, southeastern, and southwestern regions of the KMC witnessing the most intense urban growth. A strong positive correlation was established between population growth and built-up area expansion, with a correlation coefficient of 0.92 and a coefficient of determination ($R^2 = 0.92$), underscoring that ~92% of the increase in built-up areas is attributed to demographic growth. The existence of a floating population of nearly 6 million, exceeding the permanent residential population, has placed immense pressure on urban land, necessitating the development of residential infrastructure, administrative facilities, commercial establishments, and employment spaces. Collectively, these findings confirm that both residential and transient populations are the key driving forces behind the rapid and extensive urban expansion observed in Kolkata.

2.3.2.6 Spatio-Temporal Evaluation of Green Space

Over the span of three decades (1990–2020), green spaces within the Kolkata Municipal Corporation (KMC) have undergone profound transformation, witnessing an approximate reduction of 23.38 square kilometers, which represents a loss of nearly

29.23% of the total green space area (Table 2.5). This significant decline is most evident in boroughs 7, 10, 11, 12, 13, 14, 15, and 16 (Figs. 2.6 and 2.12), which have experienced severe degradation due to intensified human interventions and expanding construction activities. In 1990, these boroughs were predominantly characterized by dense and healthy vegetation, but over time, large-scale urban development and infrastructure expansion have substantially altered their natural landscape. Other boroughs, particularly those in the northern and central regions, had already witnessed widespread construction prior to 1990, leaving limited green space even before the study period began. Over the past 30 years, both the quality and quantity of green cover have diminished, as evidenced by the decline in vegetation density and health indicated in Fig. 2.12. Rapid urbanization, driven by population pressure, migration, infrastructure growth, and land use changes, has not only reduced the extent of green cover but has also fragmented the remaining patches across the city.

Urban green spaces, often regarded as the "lungs of the city," are essential for maintaining ecological balance and ensuring environmental sustainability. However, in Kolkata a megacity and the capital of West Bengal, these spaces have been under tremendous stress due to the influx of migrants and commuters (Table 2.1) and continuous urban sprawl. The study employs geospatial technologies and statistical methods to analyze the distribution, condition, and temporal transformation of green spaces in the KMC over the 30-year period. The results highlight an alarming trend, posing a major challenge for urban planners, policymakers, and administrators, as the diminishing green cover threatens biodiversity, air quality, and urban livability. The chapter concludes with a strong recommendation for immediate conservation, restoration, and sustainable management of urban green spaces. The use of maps, diagrams, and satellite-derived data provides valuable tools to guide eco-centric urban planning and policy formulation, ensuring that Kolkata's future development aligns with the principles of environmental resilience and sustainability.

2.3.3 *Policy Implications for Sustainable Land Use Management*

The findings from the spatio-temporal analysis of land use and land cover (LULC) dynamics in the Kolkata Municipal Corporation (KMC) underscore the urgent need for sustainable land use management policies that balance urban growth with environmental conservation. Rapid urban expansion, driven by population pressure, economic growth, and infrastructural development, has resulted in the depletion of green spaces, encroachment on water bodies, and the transformation of fallow lands into built-up areas. To mitigate these impacts, an integrated land management approach is essential—one that aligns urban development with ecological sustainability.

A key policy priority should be the implementation of strict land use zoning regulations, ensuring the preservation of ecologically sensitive areas such as wetlands,

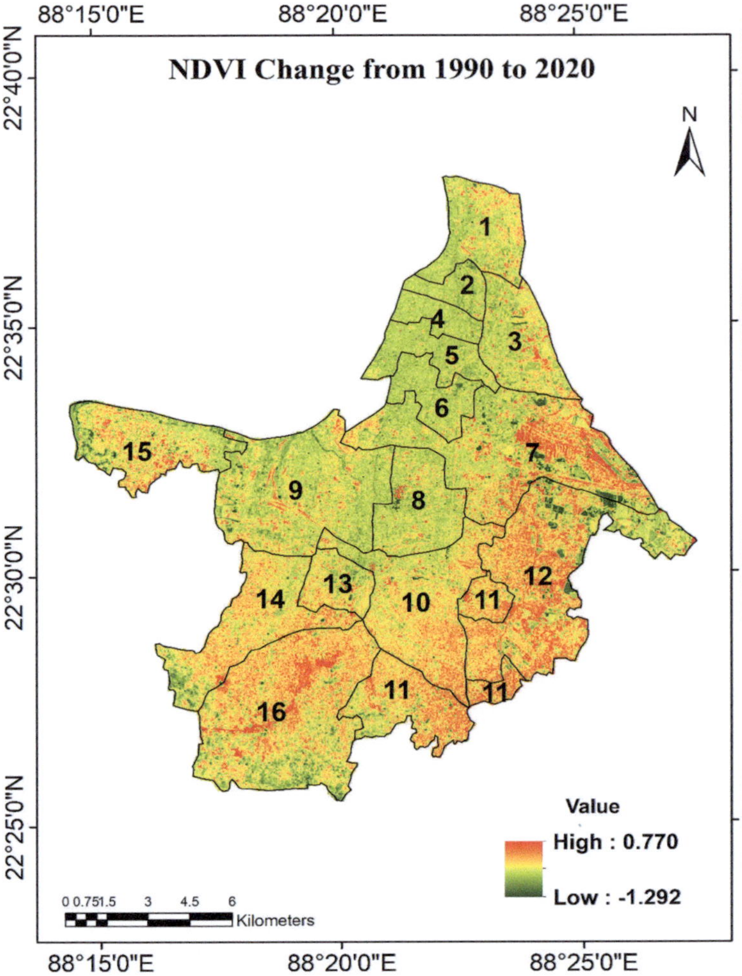

Fig. 2.12 Transformation in Vegetation Index over the last three decades

riverbanks, and urban green spaces. The East Kolkata Wetlands (Conservation and Management) Act (2006) serves as an exemplary model that could be expanded to other vulnerable zones within the city. Moreover, urban growth boundaries (UGBs) and green buffer zones should be established to limit unchecked horizontal expansion and promote vertical urban growth. The incorporation of Geospatial Information Systems (GIS) and remote sensing (RS) tools into land use planning can facilitate real-time monitoring and enforcement of land policies.

Additionally, community-based green initiatives, such as neighborhood parks, rooftop gardens, and urban forestry programs, should be promoted to enhance environmental resilience and public participation in conservation efforts. Urban renewal projects must adopt a "green infrastructure" framework that integrates ecological corridors, sustainable drainage systems, and open spaces into city planning. Furthermore, policy measures should emphasize environmental impact assessments (EIA) before approving any major construction or redevelopment projects. In conclusion, sustainable land use management in KMC requires a multi-dimensional policy framework that combines scientific monitoring, participatory governance, and eco-centric urban planning. Such policies will ensure that Kolkata evolves as a resilient, inclusive, and environmentally sustainable megacity.

Chapter 3
The Role of Urban Green Spaces in Thermal Variability and Climate Regulation

3.1 Background

Over the past three decades, urbanization has accelerated dramatically across the globe, resulting in more than half of the world's population now residing in urban areas (Moghadam & Helbich, 2013, UNDESA, 2012). This proportion is expected to surpass 60% by 2030 (Demographia, 2017) and reach 72% by 2050 (UN, 2012). Such rapid urban growth particularly evident in India has led to the steady depletion of natural resources, especially urban green spaces (Mushore et al., 2017; Sannigrahi et al., 2018; The World Resources Institute, 1996; Xian et al., 2005). Increasing population and spatial expansion have caused continuous shrinkage of urban greenery, a trend documented globally and in India (Chadchan & Shankar, 2012). Concurrently, agricultural lands, particularly on city peripheries, are being converted to urban settlements (Pandey & Seto, 2015), thereby degrading agricultural and forested areas, increasing barren lands, and enhancing impermeable surface coverage (Kumar et al., 2012). These transformations have further contributed to the rise in land surface temperatures (LST) (Shahfahad et al., 2021).

India has witnessed a notable surge in urban expansion in recent decades, with the number of million-plus cities rising from 23 in 1991 to 53 in 2011, while the urban population grew from 109 million in 1971 to 377 million in 2011, accounting for 19.9 and 31.6% of the total population, respectively (Census, 2011). Several studies have confirmed that this rapid urbanization has caused a considerable decline in urban green spaces (Chudnovsky et al., 2004; Mushore et al., 2017; Sannigrahi et al., 2018; Yogesh et al., 2009). Researchers have established strong correlations among land surface temperature, urbanized land, and green cover, linking these to the urban heat island (UHI) effect (Badarinath et al., 2005; Mallick et al., 2008). Increased LSTs contribute to the development of UHIs, altering urban microclimates (Feng et al., 2014; Rinner & Hussain, 2011) and intensifying greenhouse gas concentrations (Hartmann et al., 2013; Loehle & Scafetta, 2011), energy–water

© The Author(s), under exclusive license to Springer Nature Switzerland AG 2026
M. B. Ali and S. Jamal, *Urban Green Spaces and Environmental Sustainability*,
https://doi.org/10.1007/978-3-032-15322-7_3

imbalances (Nayak & Mandal, 2012; Sertel et al., 2011; Ward et al., 2014), and heat-wave incidences, while deteriorating air quality (Chudnovsky et al., 2004). The loss of vegetation further reduces carbon sequestration and evapotranspiration, diminishes the cooling effect, and undermines biodiversity (Karuppannan et al., 2014; McKinney, 2002). These interlinked processes significantly compromise ecosystem sustainability (Hao et al., 2016).

Urban green spaces are thus vital indicators of urban environmental health (Carrus et al., 2015), providing a multitude of ecological services such as climate regulation, air and water purification, flood control, biodiversity conservation, and recreational benefits (Escobedo et al., 2011; Gairola & Noresah, 2010; Konijnendijk et al., 2013; Nowak et al., 2010; Wolch et al., 2014; Zhang et al., 2014). The assessment of LST and UHI phenomena relies heavily on remote sensing (RS) and geographic information systems (GIS) (Deng & Wu, 2012; Ma et al., 2010; Xu et al., 2013), enabling detection of long-term spatial–temporal trends (Li & Yang, 2004; Mass, 1999; Mallick et al., 2013; Mathew et al., 2016; Salman, 2004; Yuan et al., 2005). Indicators such as LST, NDVI, and NDBI are widely applied to monitor urban ecosystem transformations (Asgarian et al., 2015; Guo et al., 2015; Mohan & Kandya, 2015; Rotem-Mindali et al., 2015), as they effectively quantify correlations between vegetation, built-up density, and temperature variations (Bindi et al., 2009; Buyadi et al., 2013; Feng et al., 2014; Jamal et al., 2023; Shahfahad et al., 2021).

Multi-temporal Landsat data provide valuable insights into surface temperature variations over decadal intervals (Buyadi et al., 2013; Chen et al., 2002; Weng, 2001), with thermal bands offering reliable LST estimation for urban analysis. Studies consistently demonstrate that built-up and industrial zones exhibit higher surface temperatures than vegetated areas, whereas increased green cover exerts a cooling influence on surrounding built-up environments (Chen et al., 2006; Hang and Rahman, 2018; Li and Yang, 2004; Yuan and Bauer, 2007). Therefore, this chapter aims to analyze the intricate relationship between built-up intensity, green cover distribution, and land surface temperature variations in the KMC area to better understand urban thermal dynamics.

3.2 Database and Methodology

3.2.1 Database

Satellite remote sensing and Geographic Information System (GIS) technologies are globally recognized as reliable and effective tools for assessing and monitoring urban green spaces, built-up expansion, and microclimatic variations, owing to their extensive spatial coverage and periodic data availability (Smith et al., 1998). In the present study, multi-temporal Landsat satellite datasets were employed to analyze temporal changes. Specifically, Landsat-5 TM imagery from 1990, 2000, and 2010, and Landsat-8 OLI imagery from 2020 were utilized. All satellite data were sourced from the USGS Earth Explorer web portal (https://earthexplorer.usgs.gov), with careful selection of cloud-free and haze-free images to ensure data precision. During the preprocessing stage, all images were georeferenced and rectified using the Universal Transverse Mercator (UTM) Zone 45 N projection, based on the WGS 84 datum. Image registration was performed through the nearest neighbor resampling technique, supported by 25 Ground Control Points (GCPs) to achieve optimal overlay accuracy. Each image maintained a spatial resolution of 30 m, following a consistent path and row configuration of 138 and 44, respectively, which ensures comparability across temporal datasets. Supplementary data, including topographical and administrative maps, were also utilized. Furthermore, Google Earth Pro served as a vital tool for collecting GCP samples and validating spatial features, particularly for linking green spaces, built-up zones, and surface temperature data.

3.2.2 Methods

Data processing and analysis were conducted using ArcGIS 10.5 software to delineate and extract the Area of Interest (AOI), based on the most recent Kolkata Municipal Corporation (KMC) boundary map. Key environmental indices namely the Normalized Difference Built-up Index (NDBI), Normalized Difference Vegetation Index (NDVI), and Land Surface Temperature (LST) were computed for the years 1990, 2000, 2010, and 2020. In addition, cross-sectional surface temperature profiles were created to compare thermal variations across the same geographic locations at decadal intervals over the 30-year period (Fig. 3.1).

To gain deeper spatial insights, a borough-wise analysis was undertaken, wherein NDBI, NDVI, and LST values were extracted using the clip tool in ArcGIS 10.5. Statistical analysis was subsequently performed using the Statistical Package for the Social Sciences (SPSS) to explore correlations between built-up intensity, vegetation cover, and land surface temperature. Finally, scatter plots were generated to visually depict the relationships among these variables, as illustrated in Figs. 3.12 and 3.13, thereby enabling a clearer understanding of the thermal and ecological dynamics within the Kolkata Municipal Corporation.

3.2.2.1 Calculation of Normalized Difference Vegetation Index

The Normalized Difference Vegetation Index (NDVI) serves as one of the most widely used indicators for assessing vegetation health and density. It is computed using Eq. 1, which utilizes the red and near-infrared (NIR) spectral bands derived from Landsat satellite data. The rationale for using these two specific bands lies in the fact that chlorophyll pigments in vegetation strongly absorb radiation in the red region while reflecting significantly in the NIR portion of the electromagnetic spectrum (Bindi et al., 2009). This spectral contrast enables effective discrimination between vegetated and non-vegetated surfaces. The NDVI formula is expressed as:

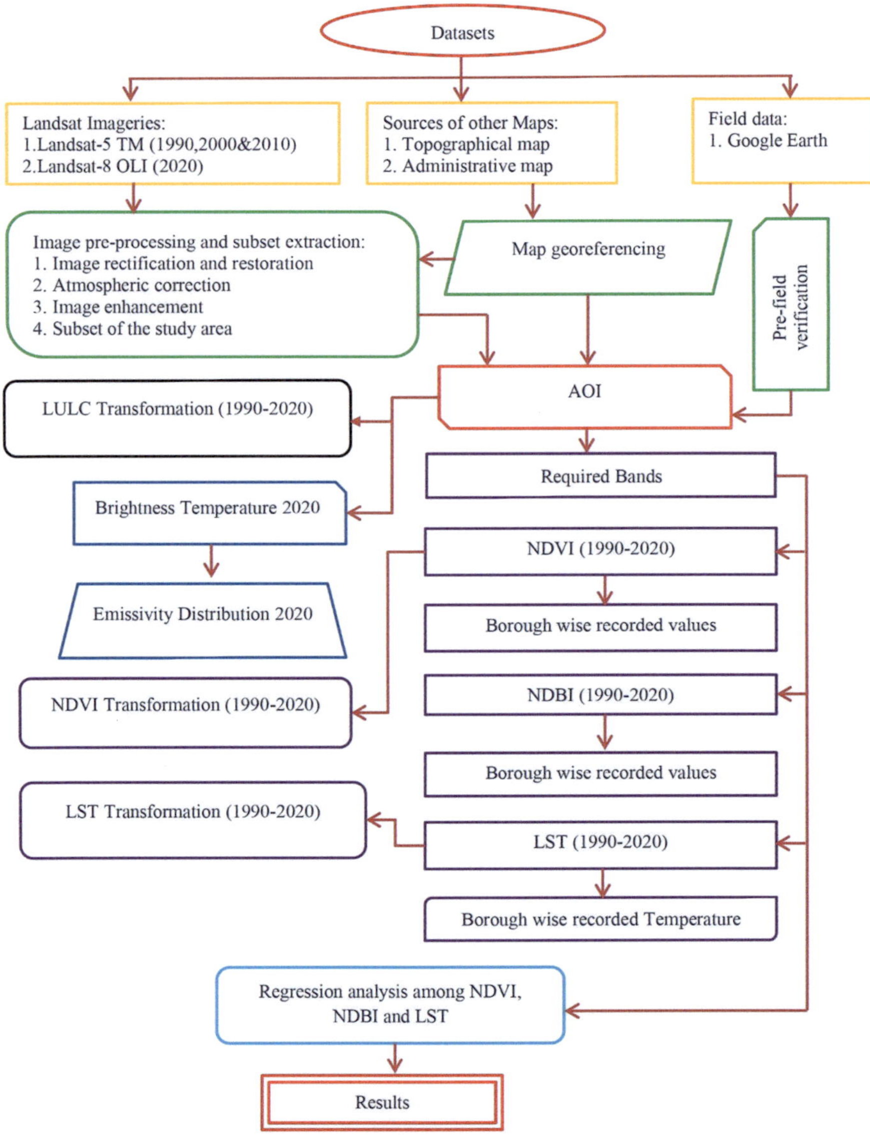

Fig. 3.1 Methodological flowchart of the chapter

$$\text{NDVI} = \frac{(\text{NIR} - \text{Red})}{(\text{NIR} + \text{Red})}.$$

This index yields values ranging between −1 and +1, where higher positive values represent dense, healthy vegetation, whereas lower or negative values correspond to non-vegetated or barren surfaces.

3.2.2.2 Calculation of Normalized Difference Built-Up Index

The Normalized Difference Built-up Index (NDBI) is used to identify built-up or impervious urban areas by utilizing the shortwave infrared (SWIR) and near-infrared (NIR) bands of Landsat imagery. The computation follows Eq. 2, developed by Zha et al. (2003). Built-up areas typically reflect more strongly in the SWIR band and less in the NIR band, making this index particularly effective in detecting urban structures. The NDBI is calculated using the following expression:

$$\mathrm{NDBI} = \frac{(\mathrm{SWIR} - \mathrm{NIR})}{(\mathrm{SWIR} + \mathrm{NIR})}.$$

Positive NDBI values indicate built-up surfaces such as buildings and roads, whereas negative values signify vegetation, water bodies, or bare land.

3.2.2.3 Calculation of LST from Thermal Band of Landsat 5 TM

The Land Surface Temperature (LST) for Landsat 5 TM data is derived through a series of systematic computational steps (Fig. 3.2). Initially, the digital number (DN) of the thermal band is converted into spectral radiance ($L\lambda$) using Eq. 3, which is based on Planck's radiation law:

$$L(\lambda) = \frac{2hc^2}{\lambda^5\left(\exp\left(\frac{hc}{\lambda kT}\right) - 1\right)}.$$

Here, $L(\lambda)$ denotes spectral radiance at wavelength λ, h is Planck's constant, c is the speed of light, k is Boltzmann's constant, T is the temperature of the emitting surface, and λ is the wavelength of emitted radiation.

Subsequently, the spectral radiance is converted into at-satellite brightness temperature in Kelvin ($T(K)$) using Eq. 4:

$$T(K) = \frac{c^2}{2kv^2} \ln\left(\frac{2hv^3}{c^2\left(\exp\left(\frac{hv}{kT_b}\right) - 1\right)}\right).$$

In this equation, $T(K)$ represents brightness temperature in Kelvin, c is the speed of light (3×10^8 m/s), k is Boltzmann's constant (1.38×10^{-23} J/K), h is Planck's constant (6.63×10^{-34} J·s), and v denotes the frequency of emitted radiation. Calibration constants used for Landsat 5 TM thermal band are $K_1 = 1260.56$ and $K_2 = 607.66$ mW·cm^{-2}·sr^{-1}, while the effective spectral range (b) is 1.239 μm, representing the wavelength range where the sensor's response exceeds 50%.

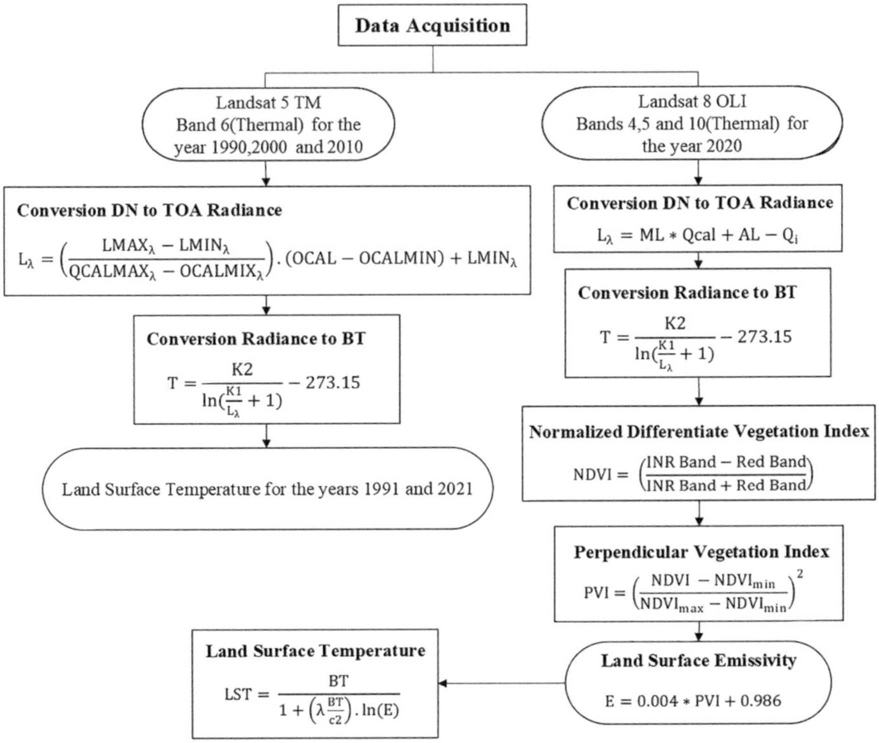

Fig. 3.2 Flowchart of land surface temperature calculation

3.2.2.4 Calculation of LST from Thermal Band of Landsat 8 OLI/TIRS

The estimation of Land Surface Temperature (LST) from Landsat 8 OLI/TIRS imagery involves a series of systematic computational steps (Fig. 3.2). Initially, the Operational Land Imager (OLI) and Thermal Infrared Sensor (TIRS) band data are converted into spectral radiance using Eq. 5, as suggested by standard USGS methodologies:

$$L_\lambda = (M_L \times Q_{\text{cal}}) + A_L.$$

Here, L_λ represents the spectral radiance measured by the sensor, M_L denotes the multiplicative rescaling factor for each band, A_L refers to the additive rescaling factor, and Q_{cal} stands for the quantized and calibrated pixel values (DN) of the standard Landsat product.

Following this, the reflectance rescaling coefficients provided in the metadata file (MTL file) are applied to convert the spectral radiance into top-of-atmosphere reflectance using Eq. 6:

$$\rho'_\lambda = M_\rho \times Q_{\mathrm{cal}} + A_\rho.$$

In this equation, ρ'_λ signifies the reflectance for a specific band (λ), M_ρ and A_ρ represent the band-specific multiplicative and additive rescaling factors, respectively, and Q_{cal} corresponds to the digital number (DN) for that band.

Conversion of Radiance to Brightness Temperature (BT)

After calculating radiance, the next step involves converting spectral radiance to brightness temperature for the TIRS band, using thermal constants available in the MTL metadata file, as given by Eq. 7:

$$\mathrm{BT} = \frac{K_2}{\ln\left(\frac{K_1}{L_\lambda} + 1\right) + \ln h}.$$

Here, BT represents the at-satellite brightness temperature (Kelvin), K_1 and K_2 are thermal calibration constants specific to the sensor, L_λ denotes spectral radiance, and $\ln h$ refers to the natural logarithm component.

Conversion of LST from Kelvin to Celsius

The brightness temperature (BT) in Kelvin is then converted into Celsius using Eq. 8, as demonstrated by Jamal et al. (2023):

$$T(^\circ\mathrm{C}) = \mathrm{BT} - 273.15.$$

Here, T(°C) denotes the land surface temperature in degrees Celsius, and BT represents the brightness temperature obtained from the TIRS band in Kelvin.

Calculation of Vegetation Proportion (PV)

To determine the proportion of vegetation, denoted as PV, a quantitative relationship is established using the Normalized Difference Vegetation Index (NDVI). The formula is expressed as:

$$\mathrm{PV} = \left[\frac{(\mathrm{NDVI} - \mathrm{NDVI}_{\min})}{(\mathrm{NDVI}_{\max} - \mathrm{NDVI}_{\min})}\right]^2.$$

In this equation, PV represents the proportion of vegetation, $\mathrm{NDVI}_{\min}$ is the minimum NDVI value, and $\mathrm{NDVI}_{\max}$ denotes the maximum NDVI value derived from the image dataset. This parameter serves as a critical input for refining surface emissivity calculations and accurately estimating LST in heterogeneous urban environments.

3.2.2.5 Regression Analysis

In the present study, a regression analysis was conducted to explore and quantify the relationship between multiple variables. To visually represent this relationship, a scatter plot was generated along with a linear regression line, illustrating the degree and direction of correlation among the observed parameters. The analysis was carried out using Microsoft Excel 2021, applying the linear regression equation as expressed below:

$$Y_c = a + b.$$

Here, Y_c denotes the predicted or estimated value, a refers to the y-intercept, and b represents the regression coefficient (slope), which indicates the rate of change in the dependent variable concerning the independent variable. The constants a and b were derived using standard statistical formulas, enabling the study to identify trends and establish the strength of relationships between selected environmental and urbanization indicators.

$$b = \frac{N\sum XY - \sum X \sum Y}{N\sum X^2 - \left(\sum X\right)^2},\quad a = \frac{\sum Y - b\sum X}{N}$$

3.3 Results and Discussion

3.3.1 Transformation in Vegetation Index and Its Association with Thermal Conditions

Table 3.1 presents the maximum and minimum values of the Normalized Difference Vegetation Index (NDVI) across different boroughs for the years 1990, 2000, 2010, and 2020. NDVI serves as an essential indicator of vegetation density and health within a particular area. For each borough, the table lists two distinct NDVI values for every year: the maximum NDVI value, representing zones with dense vegetation, and the minimum NDVI value, denoting areas with minimal or no vegetation cover. Over the 30-year period from 1990 to 2020, the data reveals a steady decline in both dense and sparse vegetation categories, while regions classified under other land use and cover types have expanded significantly. This persistent reduction in dense vegetation cover is one of the major factors contributing to the increase in land surface temperature (LST). Furthermore, green spaces with sparse vegetation, as identified through NDVI values, have also experienced noticeable degradation over time.

To gain deeper insight, a ward-level analysis of NDVI indices has been carried out, enabling a more localized understanding of vegetation dynamics. This spatially

Table 3.1 Borough-wise recorded highest and lowest NDVI for 2020, 2010, 2000 and 1990

Boroughs	2020		2010		2000		1990	
	Highest	Lowest	Highest	Lowest	Highest	Lowest	Highest	Lowest
1	0.467	−0.047	0.527	−0.217	0.417	−0.170	0.655	−0.474
2	0.416	0.012	0.437	−0.086	0.275	−0.094	0.443	−0.476
3	0.478	−0.057	0.562	−0.189	0.469	−0.167	0.588	−0.333
4	0.378	−0.041	0.419	−0.137	0.259	−0.132	0.519	−0.278
5	0.405	−0.091	0.476	−0.200	0.406	−0.191	0.518	−0.561
6	0.479	−0.014	0.449	−0.107	0.377	−0.115	0.553	−0.446
7	0.485	−0.073	0.613	−0.209	0.496	−0.160	0.705	−0.682
8	0.423	−0.072	0.505	−0.179	0.462	−0.137	0.645	−0.690
9	0.483	−0.094	0.541	−0.375	0.477	−0.259	0.741	−0.534
10	0.480	−0.098	0.556	−0.106	0.478	−0.107	0.652	−0.563
11	0.473	−0.059	0.533	−0.156	0.473	−0.098	0.687	−0.436
12	0.496	−0.159	0.600	−0.265	0.472	−0.143	0.685	−0.836
13	0.456	−0.063	0.575	−0.163	0.504	−0.132	0.667	−0.540
14	0.556	−0.048	0.565	−0.163	0.521	−0.102	0.641	−0.632
15	0.478	−0.088	0.603	−0.375	0.472	−0.292	0.766	−0.340
16	0.556	−0.086	0.583	−0.149	0.548	−0.148	0.695	−0.333

detailed examination helps identify specific wards that have undergone substantial vegetation loss or land cover transformation. Such localized analysis is vital for assessing the environmental implications of urban growth and for formulating targeted conservation and land management interventions.

Figure 3.3 displays the NDVI variations from 1990 to 2020, highlighting significant inter-borough differences. For example, in borough 1, the highest NDVI recorded in 2020 was 0.467, while the lowest was −0.047, indicating a broad range of vegetation density. Similarly, borough 7 exhibited a gradual decline in its maximum NDVI value from 0.705 in 1990 to 0.613 in 2010, and further down to 0.485 in 2020. Its minimum NDVI value, however, improved from −0.682 in 1990 to −0.073 in 2020, suggesting a reduction in barren or impervious surfaces.

Highly urbanized boroughs tend to exhibit lower NDVI values, primarily due to the dominance of impervious surfaces such as roads, buildings, and infrastructure that restrict vegetation growth. In contrast, boroughs with lower urban intensity generally record higher NDVI values, reflecting greater vegetation cover and open green spaces. In certain cases, boroughs with relatively stable or increasing NDVI values may have benefited from proactive conservation measures, such as afforestation, reforestation, or urban greening programs, which have helped sustain or enhance vegetation cover.

The spatial transformation portrayed in Fig. 3.4 highlights substantial alterations in boroughs 11, 12, and 16, where marked temperature increases between 1990

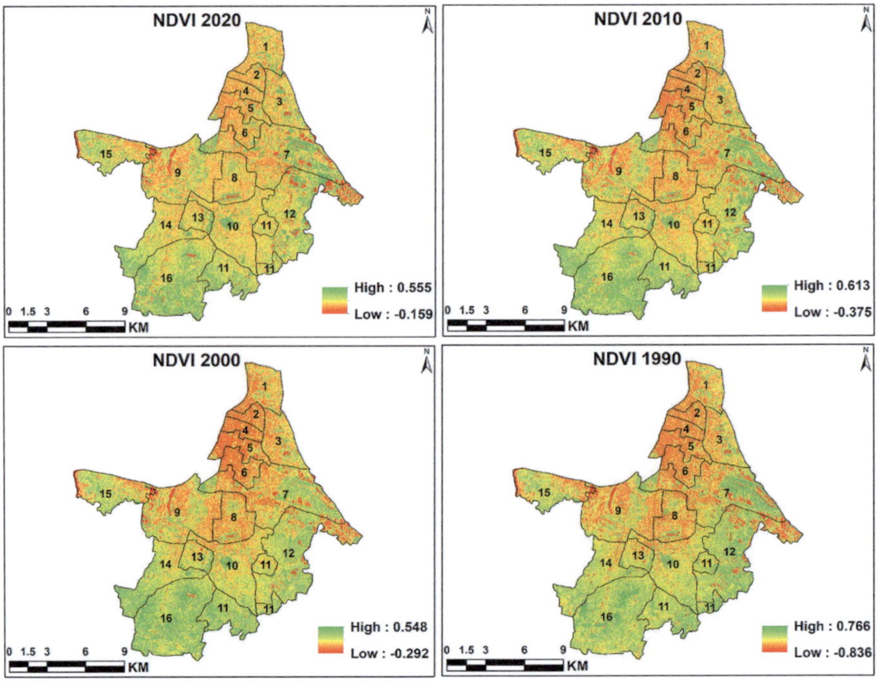

Fig. 3.3 Normalized difference vegetation index (1990–2020)

and 2020 coincide with a pronounced reduction in green cover. These observations underscore a progressive decline in NDVI values within the urban core over time. Numerous prior studies have confirmed a strong inverse correlation between NDVI and LST, indicating that areas with diminished vegetation cover experience higher surface temperatures, as evidenced by significant coefficients of determination (Fig. 3.12).

3.3.2 Transformation in Built-Up Index and Its Association with Thermal Conditions

The Normalized Difference Built-up Index (NDBI) serves as a key remote sensing indicator used to delineate and analyze urbanized or built-up areas. Table 3.2 presents the highest and lowest NDBI values recorded across different boroughs of the Kolkata Municipal Corporation (KMC) for the years 1990, 2000, 2010, and 2020. Each borough entry comprises two values per year: the maximum and minimum NDBI values, signifying areas with the highest and lowest built-up intensities, respectively. For instance, in 2020, borough 1 recorded a maximum NDBI of 0.139 and a minimum of −0.311, whereas in 1990, the maximum and minimum NDBI values for the same

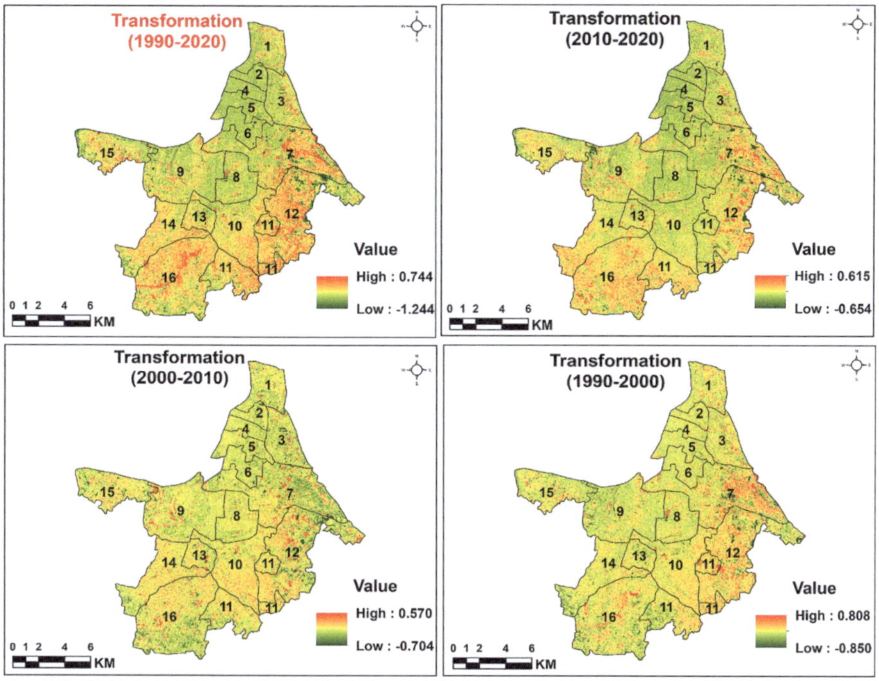

Fig. 3.4 Maps of decadal and overall transformation in normalized difference vegetation index (1990–2020)

borough were 0.603 and −0.634, respectively. These figures highlight the changing spatial extent and intensity of urbanization over the last three decades. Generally, positive NDBI values denote regions with dense built-up structures, while negative values indicate areas with sparse or no built-up features. The dataset offers a robust foundation for analyzing urban growth dynamics, assessing spatial transformations, and guiding urban planning and management strategies for sustainable development.

Several studies have explored the intricate linkages between urbanization, population density, industrial expansion, and elevated surface temperatures, confirming that densely populated and heavily industrialized zones exhibit higher Land Surface Temperatures (LST) (Lillesand et al. 2015; Su et al. 2012; Shahfahad et al. 2021). These researchers employed Landsat TM/ETM + imagery, integrating both NDVI and NDBI indices to investigate the urban heat island (UHI) phenomenon. The findings reveal a direct and positive correlation between the proliferation of impervious surfaces and elevated surface temperatures. Moreover, other scholars have emphasized that the reduction in vegetative cover significantly amplifies the UHI effect (Weng et al. 2004; Xiao et al. 2008).

Figure 3.5 depicts the spatial distribution and expansion of impervious built-up areas during 1990, 2000, 2010, and 2020, categorized by NDBI values. The maps reveal a consistent rise in both the spatial extent and intensity of built-up land

Table 3.2 Borough-wise recorded highest and lowest NDBI for 2020, 2010, 2000, and 1990

Boroughs	2020		2010		2000		1990	
	Highest	Lowest	Highest	Lowest	Highest	Lowest	Highest	Lowest
1	0.139	−0.311	0.425	−0.290	0.400	−0.250	0.603	−0.634
2	0.124	−0.253	0.372	−0.120	0.347	−0.093	0.371	−0.135
3	0.169	−0.290	0.550	−0.355	0.389	−0.368	0.648	−0.545
4	0.105	−0.222	0.369	−0.200	0.322	−0.125	0.603	−0.386
5	0.168	−0.230	0.619	−0.200	0.333	−0.235	0.617	−0.343
6	0.177	−0.329	0.388	−0.186	0.346	−0.127	0.411	−0.429
7	0.201	−0.380	0.431	−0.600	0.530	−0.429	0.759	−0.963
8	0.157	−0.278	0.387	−0.294	0.408	−0.286	0.708	−0.441
9	0.187	−0.350	0.487	−0.355	0.463	−0.353	0.667	−0.966
10	0.215	−0.321	0.487	−0.231	0.431	−0.231	0.646	−0.625
11	0.132	−0.356	0.400	−0.353	0.350	−0.333	0.571	−0.449
12	0.167	−0.369	0.394	−0.586	0.435	−0.563	0.694	−0.882
13	0.126	−0.327	0.383	−0.235	0.364	−0.280	0.437	−0.385
14	0.141	−0.388	0.397	−0.300	0.351	−0.313	0.375	−0.429
15	0.134	−0.332	0.431	−0.478	0.424	−0.290	0.713	−0.565
16	0.155	−0.357	0.424	−0.422	0.431	−0.389	0.459	−0.596

over time, largely driven by rapid population growth, infrastructure development, and urban expansion. Furthermore, Fig. 3.13 illustrates the relationship between NDBI and LST over the same period, incorporating linear regression models for each decade. The coefficients of determination (R^2 values) 0.4896 (1990), 0.542 (2000), 0.5455 (2010), and 0.4773 (2020) collectively indicate a strong positive relationship between built-up intensity and land surface temperature, confirming that the increase in impervious surfaces has significantly contributed to the thermal escalation within the Kolkata Municipal Corporation (Fig. 3.13).

3.3.3 Spatio-Temporal Dynamism of Thermal Conditions

Land Surface Temperature (LST) represents the temperature of the Earth's surface as derived from satellite data and serves as a crucial indicator for assessing urban thermal environments. Table 3.3 and Fig. 3.6 illustrate the highest and lowest LST values recorded across the boroughs of the Kolkata Municipal Corporation (KMC) for the years 1990, 2000, 2010, and 2020. Each borough's row corresponds to its respective data, while the columns denote the highest and lowest LST values for each time period, measured in degrees Celsius. For instance, in borough 1, the highest recorded surface temperature in 2020 was 31.29 °C, and the lowest was 24.16 °C,

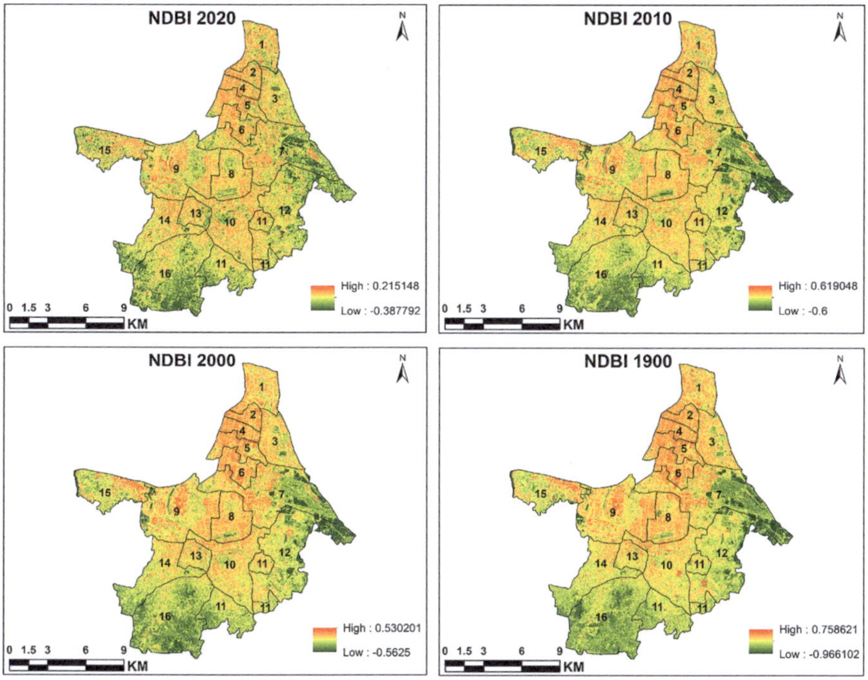

Fig. 3.5 Normalized difference built-up index (1990–2020)

indicating a clear thermal variation within the borough. These values reflect the temporal and spatial fluctuations in LST across different administrative regions of KMC (Table 3.3; Fig. 3.9).

An analysis of the tabulated data reveals that over the three-decade period, there has been a consistent rise in mean surface temperatures, highlighting a gradual intensification of the urban heat island (UHI) phenomenon. The increase in impervious surfaces such as concrete pavements, buildings, and industrial zones coupled with the decline of green cover and water bodies, has contributed significantly to heat accumulation in urban cores. As a result, the central parts of Kolkata have become warmer than peripheral areas (Fig. 3.6). This temperature disparity between urban and rural zones underscores the expanding influence of UHI, which is strongly linked to the city's unrestrained urbanization.

Table 3.3 indicates that the overall temperature range within KMC lies between 20 and 35 °C, with mean decadal values fluctuating between 21.6 and 26 °C (Fig. 3.8). Spatial analysis shows that boroughs have experienced differing levels of temperature variation across decades (Fig. 3.7). For example, borough 1 exhibited an increase in its maximum surface temperature from 30.01 °C in 1990 to 31.29 °C in 2020, while its minimum temperature rose from 21.06 to 24.16 °C during the same period. Similarly, borough 7 consistently registered the highest LST values among all boroughs, emphasizing localized heat concentration zones.

Table 3.3 Borough-wise recorded highest and lowest LST for 2020, 2010, 2000 and 1990

Boroughs	2020		2010		2000		1990	
	Highest	Lowest	Highest	Lowest	Highest	Lowest	Highest	Lowest
1	31.29	24.16	31.65	24.97	32.46	24.55	30.01	21.06
2	30.22	25.48	31.24	26.25	31.24	25.83	25.83	22.82
3	30.80	22.79	31.24	24.11	31.24	23.68	27.93	18.83
4	30.11	25.76	31.24	25.83	30.83	25.83	25.83	22.82
5	30.74	24.00	31.24	25.40	31.65	24.55	27.93	21.06
6	30.22	24.39	30.83	24.55	30.42	24.55	27.52	23.25
7	32.83	21.43	33.70	22.38	31.24	22.82	27.52	13.31
8	30.22	22.77	31.65	24.11	30.01	23.68	27.10	18.83
9	33.00	22.29	32.86	23.25	32.46	23.68	27.93	18.83
10	30.47	24.01	30.83	24.55	31.24	24.11	26.67	18.83
11	27.71	22.44	27.93	23.25	27.10	23.68	23.68	18.83
12	29.79	20.90	29.18	22.40	27.93	22.82	24.55	10.90
13	29.50	23.48	30.01	24.55	29.18	24.97	23.68	20.18
14	28.59	22.66	28.77	23.25	29.60	23.25	23.68	19.73
15	31.74	22.86	32.05	24.11	32.86	23.68	28.35	19.73
16	28.77	22.20	28.77	23.25	28.35	22.38	24.11	17.93

Several key factors contribute to these variations in surface temperature. The urban heat island effect, combined with changes in land use and land cover, plays a dominant role. Increased built-up density, reduced vegetation, and extensive concrete and asphalt surfaces lead to greater absorption and retention of solar radiation. Additionally, climatic and meteorological parameters such as humidity, wind speed, and cloud cover also influence the magnitude of LST. Conversely, boroughs that have adopted heat mitigation measures, including the development of urban green spaces, rooftop vegetation, and reflective construction materials, display relatively lower temperatures.

The correlation between NDVI and LST has been further validated through the analysis of ground control point (GCP) samples (Fig. 3.11) and regression models (Fig. 3.12), demonstrating a strong inverse relationship between vegetation cover and surface temperature. However, while Table 3.3 provides a detailed borough-wise comparison, additional temporal and climatic data are required to establish more precise trends and causative patterns in the observed thermal dynamics of the Kolkata Municipal Corporation (Figs. 3.7, 3.8, and 3.9).

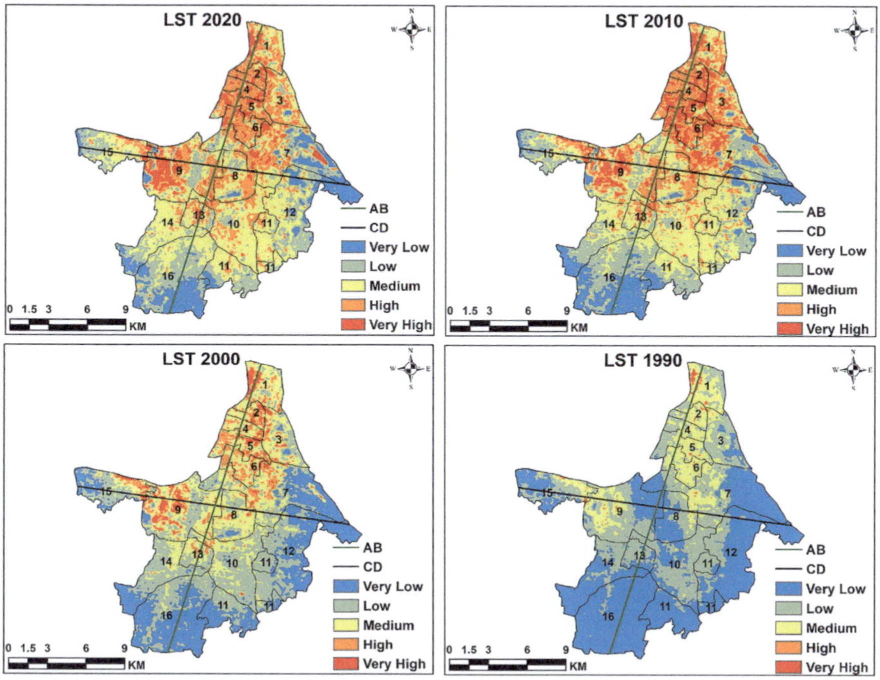

Fig. 3.6 Decadal maps of surface temperature (1990–2020)

3.3.4 Types of Land Use and Their Associations with Surface Temperature

Areas rich in vegetation, such as Maidan, Golf Garden, and the Calcutta Golf Club, exhibit noticeably lower surface temperatures compared to densely urbanized zones (Fig. 3.10). The Normalized Difference Vegetation Index (NDVI) effectively captures vegetation density and vigor, which play a vital role in moderating urban temperatures. Vegetation mitigates heat through processes like evaporative cooling, transpiration, and shading, all of which reduce surface heat absorption and ambient temperature. Furthermore, the presence of green cover enhances air circulation and ventilation, allowing trapped heat to dissipate more efficiently and contributing to localized cooling effects. These areas also display low built-up index values, signifying a sparse concentration of man-made structures. The reduced density of impervious surfaces, such as concrete and asphalt, limits heat retention and promotes thermal balance, thereby contributing to a decrease in Land Surface Temperature (LST).

In contrast, highly urbanized and densely constructed regions, such as Sobhabazar, Tikiapara, and the Dhapa dumping ground area, register significantly higher surface temperatures (Fig. 3.10). The primary reason behind this thermal increase is the extensive presence of impervious materials, including concrete and asphalt, which

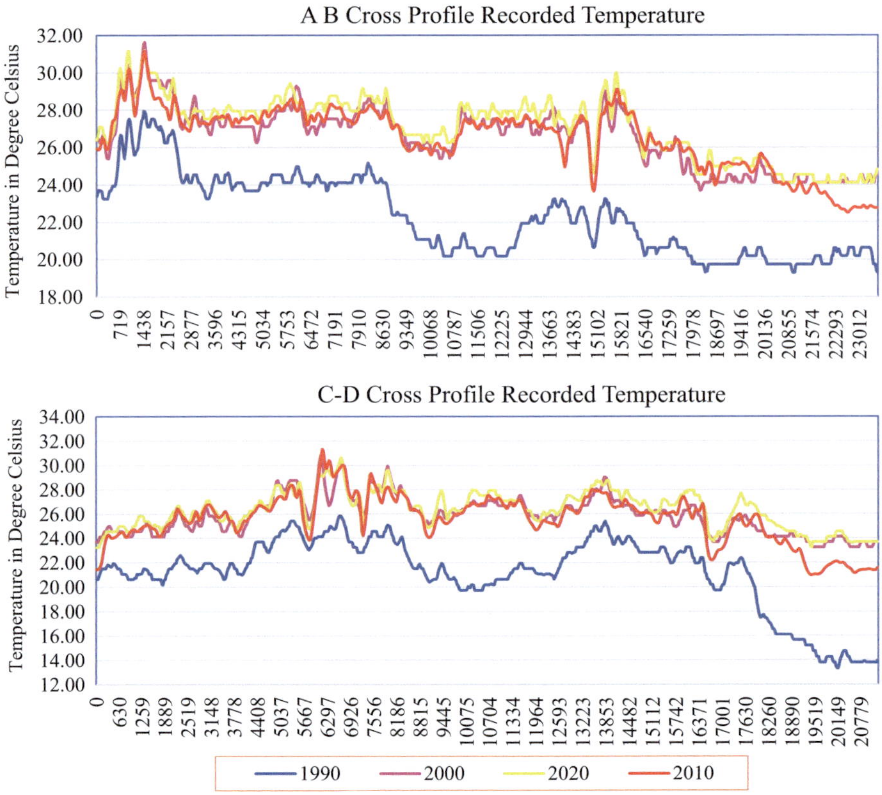

Fig. 3.7 Cross-profile recorded temperature for the years 1990, 2000, 2010, and 2020

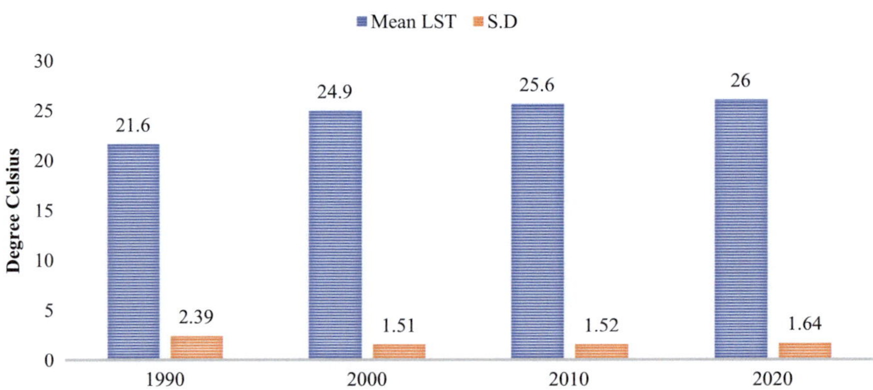

Fig. 3.8 Mean land surface temperature (1990–2020)

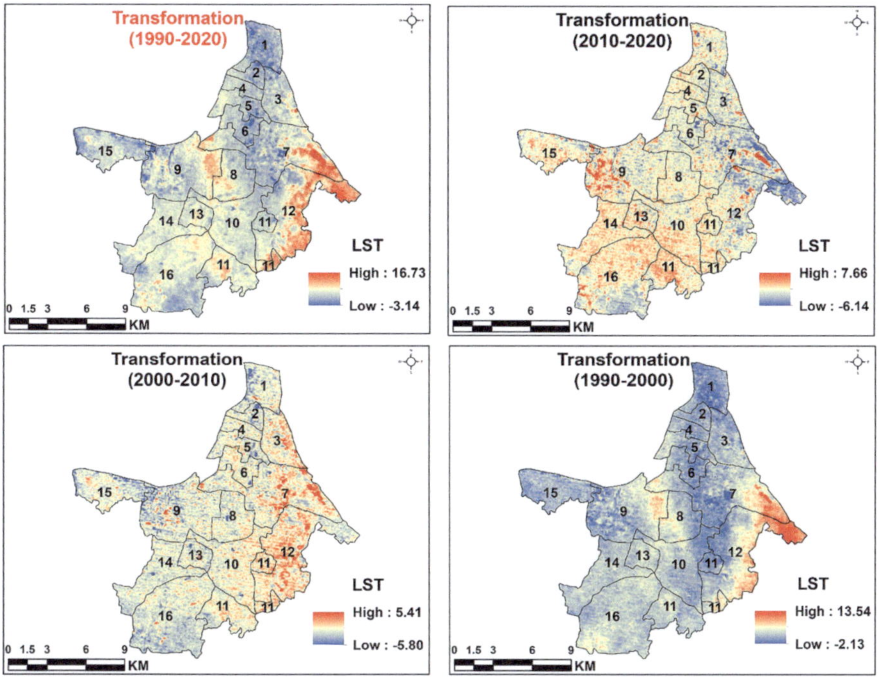

Fig. 3.9 Spatio-temporal transformation of surface temperature (1990–2020)

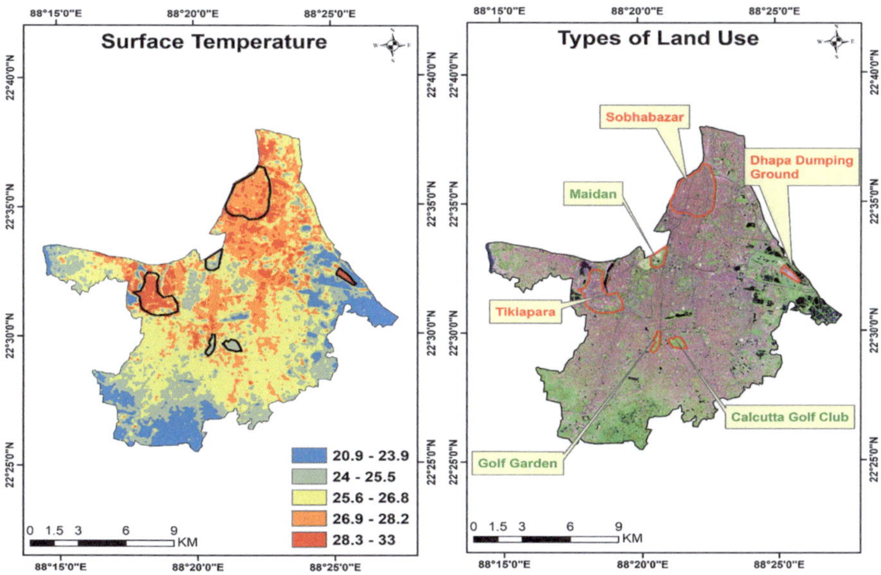

Fig. 3.10 Types of land use and recorded surface temperature

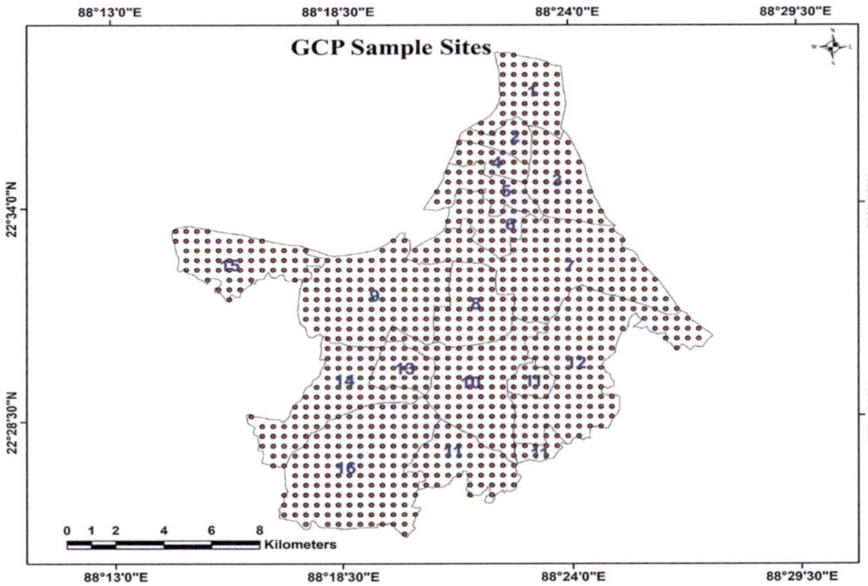

Fig. 3.11 GCP sample sites

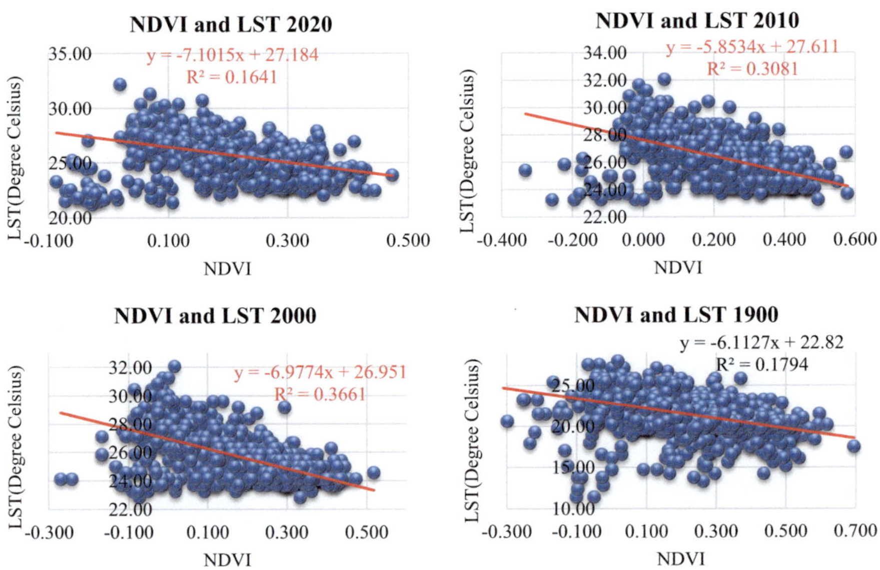

Fig. 3.12 Relationship between vegetation index and surface temperature

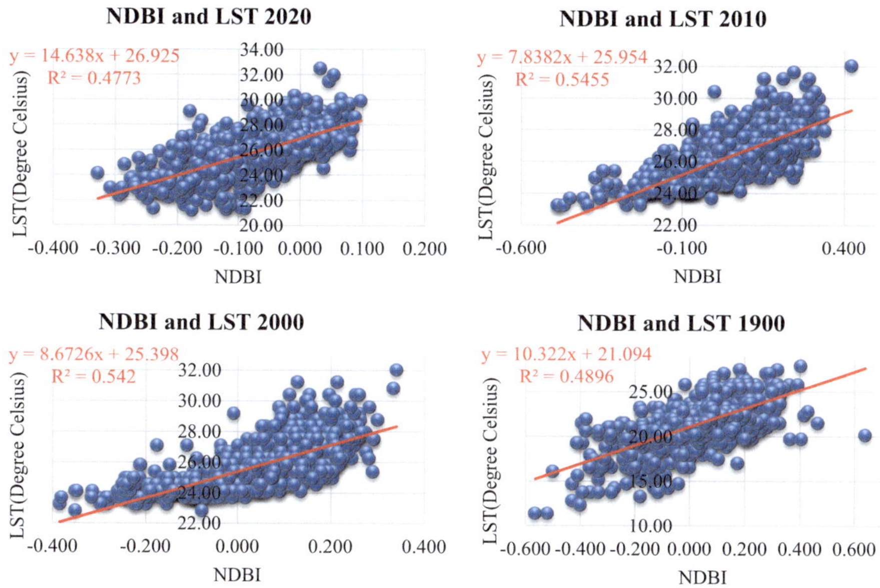

Fig. 3.13 Relationship between built-up index and surface temperature

absorb and retain solar radiation during the day and re-emit it as heat at night. The dominance of built-up infrastructure in these locations, coupled with limited vegetation and inadequate ventilation, exacerbates heat accumulation and contributes to the urban heat island (UHI) phenomenon. The concentration of buildings, industrial units, and transport networks further intensifies local warming effects, resulting in elevated LST values.

Overall, this spatial pattern underscores a clear relationship between land use type and surface temperature within the Kolkata Municipal Corporation area: regions with dense vegetation act as natural cool zones, while heavily built-up urban cores function as thermal hotspots, reinforcing the environmental importance of maintaining and expanding urban green spaces for sustainable temperature regulation.

3.3.5 Association Among Vegetation Index, Built-Up Index and Thermal Conditions

An examination of Fig. 3.6 and the accompanying graphical analysis reveals a steady and notable increase in Land Surface Temperature (LST) across the Kolkata Municipal Corporation (KMC) between 1990 and 2020. The figure clearly depicts a progressive warming trend over the three decades. In 1990, elevated temperatures were concentrated in small pockets primarily located in the western sections of the city. By 2000, these high-temperature zones had expanded, encompassing a greater number of wards on both sides of the KMC area. This expansion intensified by 2010, with thermal hotspots spreading further, and by 2020, a substantial portion of the city's wards exhibited significantly elevated temperatures. This pronounced increase in surface temperature directly correlates with the rise in built-up areas and the decline of green cover, as evidenced by variations in NDVI values over the study period (Tables 3.4, 3.5, 3.6, and 3.7).

The cross-sectional profiles of LST illustrated in Fig. 3.7 further emphasize marked fluctuations in surface temperatures across different decades, affirming the spatial–temporal transformation of the city's thermal landscape. This study establishes a strong inverse relationship between green cover and surface temperature, highlighting the critical cooling role played by vegetation. Green cover mitigates heat by facilitating shade, transpiration, and evapotranspiration, which together reduce surface heat absorption and lower ambient temperatures (Hang & Rahman, 2018). The persistent decline in vegetation, coupled with rising LST, signals an alarming trend that threatens ecological stability (Morton, 1983; Rosenberg et al., 1983; Rai, 2016). The ongoing loss of green spaces can largely be attributed to rapid population growth, urban expansion, and intensive construction activity, particularly in metropolitan regions like Kolkata (Owen et al., 2006; Sieghardt et al., 2005).

Tables 3.4, 3.5, 3.6, and 3.7 present borough-wise highest and lowest recorded values of NDVI, NDBI, and LST for the years 1990, 2000, 2010, and 2020, clearly depicting the interplay between vegetation density, built-up intensity, and surface

Table 3.4 Borough-wise recorded highest and lowest NDVI, NDBI, and LST for 1990

Boroughs	NDVI		NDBI		LST	
	Highest	Lowest	Highest	Lowest	Highest	Lowest
1	0.655	−0.474	0.603	−0.634	30.01	21.06
2	0.443	−0.476	0.371	−0.135	25.83	22.82
3	0.588	−0.333	0.648	−0.545	27.93	18.83
4	0.519	−0.278	0.603	−0.386	25.83	22.82
5	0.518	−0.561	0.617	−0.343	27.93	21.06
6	0.553	−0.446	0.411	−0.429	27.52	23.25
7	0.705	−0.682	0.759	−0.963	27.52	13.31
8	0.645	−0.690	0.708	−0.441	27.10	18.83
9	0.741	−0.534	0.667	−0.966	27.93	18.83
10	0.652	−0.563	0.646	−0.625	26.67	18.83
11	0.687	−0.436	0.571	−0.449	23.68	18.83
12	0.685	−0.836	0.694	−0.882	24.55	10.90
13	0.667	−0.540	0.437	−0.385	23.68	20.18
14	0.641	−0.632	0.375	−0.429	23.68	19.73
15	0.766	−0.340	0.713	−0.565	28.35	19.73
16	0.695	−0.333	0.459	−0.596	24.11	17.93

Table 3.5 Borough-wise recorded highest and lowest NDVI, NDBI, and LST for 2000

Boroughs	NDVI		NDBI		LST	
	Highest	Lowest	Highest	Lowest	Highest	Lowest
1	0.417	−0.170	0.400	−0.250	32.46	24.55
2	0.275	−0.094	0.347	−0.093	31.24	25.83
3	0.469	−0.167	0.389	−0.368	31.24	23.68
4	0.259	−0.132	0.322	−0.125	30.83	25.83
5	0.406	−0.191	0.333	−0.235	31.65	24.55
6	0.377	−0.115	0.346	−0.127	30.42	24.55
7	0.496	−0.160	0.530	−0.429	31.24	22.82
8	0.462	−0.137	0.408	−0.286	30.01	23.68
9	0.477	−0.259	0.463	−0.353	32.46	23.68
10	0.478	−0.107	0.431	−0.231	31.24	24.11
11	0.473	−0.098	0.350	−0.333	27.10	23.68
12	0.472	−0.143	0.435	−0.563	27.93	22.82
13	0.504	−0.132	0.364	−0.280	29.18	24.97
14	0.521	−0.102	0.351	−0.313	29.60	23.25
15	0.472	−0.292	0.424	−0.290	32.86	23.68
16	0.548	−0.148	0.431	−0.389	28.35	22.38

Table 3.6 Borough-wise recorded highest and lowest NDVI, NDBI, and LST for 2010

Boroughs	NDVI		NDBI		LST	
	Highest	Lowest	Highest	Lowest	Highest	Lowest
1	0.527	−0.217	0.425	−0.290	31.65	24.97
2	0.437	−0.086	0.372	−0.120	31.24	26.25
3	0.562	−0.189	0.550	−0.355	31.24	24.11
4	0.419	−0.137	0.369	−0.200	31.24	25.83
5	0.476	−0.200	0.619	−0.200	31.24	25.40
6	0.449	−0.107	0.388	−0.186	30.83	24.55
7	0.613	−0.209	0.431	−0.600	33.70	22.38
8	0.505	−0.179	0.387	−0.294	31.65	24.11
9	0.541	−0.375	0.487	−0.355	32.86	23.25
10	0.556	−0.106	0.487	−0.231	30.83	24.55
11	0.533	−0.156	0.400	−0.353	27.93	23.25
12	0.600	−0.265	0.394	−0.586	29.18	22.40
13	0.575	−0.163	0.383	−0.235	30.01	24.55
14	0.565	−0.163	0.397	−0.300	28.77	23.25
15	0.603	−0.375	0.431	−0.478	32.05	24.11
16	0.583	−0.149	0.424	−0.422	28.77	23.25

Table 3.7 Borough-wise recorded highest and lowest NDVI, NDBI, and LST for 2020

Boroughs	NDVI		NDBI		LST	
	Highest	Lowest	Highest	Lowest	Highest	Lowest
1	0.467	−0.047	0.139	−0.311	31.29	24.16
2	0.416	0.012	0.124	−0.253	30.22	25.48
3	0.478	−0.057	0.169	−0.290	30.80	22.79
4	0.378	−0.041	0.105	−0.222	30.11	25.76
5	0.405	−0.091	0.168	−0.230	30.74	24.00
6	0.479	−0.014	0.177	−0.329	30.22	24.39
7	0.485	−0.073	0.201	−0.380	32.83	21.43
8	0.423	−0.072	0.157	−0.278	30.22	22.77
9	0.483	−0.094	0.187	−0.350	33.00	22.29
10	0.480	−0.098	0.215	−0.321	30.47	24.01
11	0.473	−0.059	0.132	−0.356	27.71	22.44
12	0.496	−0.159	0.167	−0.369	29.79	20.90
13	0.456	−0.063	0.126	−0.327	29.50	23.48
14	0.556	−0.048	0.141	−0.388	28.59	22.66
15	0.478	−0.088	0.134	−0.332	31.74	22.86
16	0.556	−0.086	0.155	−0.357	28.77	22.20

temperature. Areas exhibiting high NDVI values tend to have lower LST, confirming the cooling effect of vegetation, while regions with high NDBI values indicative of dense built-up structures display elevated LST values.

Green cover is thus indispensable for maintaining thermal balance in urban environments. By providing shade and facilitating evapotranspiration, vegetation significantly lowers surface and air temperatures. Integrating green infrastructure such as parks, gardens, and tree-lined streets into urban planning is essential for mitigating the urban heat island (UHI) effect and fostering sustainable urban ecosystems. However, continued urban sprawl and infrastructure development often lead to the conversion of vegetated areas into impervious surfaces, intensifying heat accumulation.

The study ultimately emphasizes the necessity of protecting and expanding urban green spaces to counteract the escalating land surface temperatures. The strong correlations observed between NDVI, NDBI, and LST (Figs. 3.11, 3.12, and 3.13) reinforce the conclusion that vegetation loss directly contributes to thermal intensification, while built-up proliferation exacerbates the warming trend. Therefore, promoting urban greening initiatives and sustainable land management strategies is crucial to restoring ecological balance and enhancing urban livability.

Chapter 4
Urban Green Spaces and Environmental Pollution Abatement

4.1 Background

The accelerated pace of economic and industrial growth in developing nations has resulted in extensive ambient air pollution, significantly affecting various aspects of society, including public health and economic productivity particularly in densely populated urban centers. Given the strong association between air pollution levels and the prevalence of numerous diseases, there has been increasing concern and attention from both governments and the public regarding urban air quality (Chen et al., 2016; Lelieveld et al., 2015; Li et al., 2015). Globally, ambient air pollution is estimated to cause over 4 million premature deaths annually, with certain regions still witnessing a continued rise in pollution levels (Shaddick et al., 2020). Prolonged exposure to pollutants greatly elevates the risks of cardiovascular and respiratory diseases as well as other health complications (Thurston et al., 2017). Of particular concern is particulate matter (PM) airborne particles of varying size and composition, which constitutes the leading environmental health risk worldwide and ranks fifth among the major global health threats (Ali et al., 2019; Cohen et al., 2017; Di et al., 2017). While reducing PM concentration is known to yield substantial public health benefits, the challenge extends beyond emission control to encompass broader air quality management (Burns et al., 2019).

Recent studies emphasize the role of vegetation and urban greenery in mitigating the effects of PM through complex physical, biological, and chemical mechanisms (Wolf et al., 2020). As urbanization intensifies, so does the awareness of the influence of natural environments on human health, prompting interest in nature-based interventions such as urban green spaces, which serve as buffers against the adverse effects of pollution. These nature-based solutions (NBS) utilize natural processes to combat environmental issues effectively and often offer long-term cost efficiency compared to technological alternatives (Bauduceau et al., 2015). In urban contexts, blue-green infrastructure comprising interconnected networks of vegetation, water bodies, and open spaces has been shown to improve air quality and enhance human

© The Author(s), under exclusive license to Springer Nature Switzerland AG 2026
M. B. Ali and S. Jamal, *Urban Green Spaces and Environmental Sustainability*,
https://doi.org/10.1007/978-3-032-15322-7_4

well-being (Liu & Shen, 2014; Tzoulas et al., 2007). However, not all green spaces perform equally; their efficiency depends on structural composition, including the presence of trees, shrubs, and herbaceous vegetation. Trees, in particular, play a crucial role by filtering pollutants through surface deposition and stomatal absorption, as well as by cooling urban air via shade and evapotranspiration (McDonald et al., 2016; Nowak et al., 2006).

Despite growing recognition of their benefits, there remains a lack of detailed understanding regarding the design, composition, and management of green spaces that most effectively enhance air purification and microclimate regulation. A spatially explicit, high-resolution approach is essential for understanding the dispersion and impact of atmospheric pollutants such as PM (Pinho et al., 2012). Moreover, the interplay between green spaces and air pollution is multifaceted, involving contributions from diverse disciplines such as atmospheric sciences, plant physiology, and urban planning (Sun & Zhu, 2019; Wolf et al., 2020). Estimating the indirect health benefits of green spaces through improved air quality remains challenging due to the complexity of causal mechanisms and multiple co-occurring health effects (Arantes & Mauad, 2019; Janhäll, 2015; Markevych et al., 2017; Vienneau et al., 2017).

Parallel to air quality, green spaces also influence urban noise reduction. Margaritis and Kang (2016) demonstrated the importance of integrating greenery with built-up and traffic-prone areas to reduce traffic noise across eight cities. Similarly, Lopez and Souza (2018) observed that native vegetation contributes to noise mitigation, even when not explicitly designed as a barrier. Other research highlights how urban structural features such as buildings, roads, and vegetation shape the spatial distribution of traffic-generated noise (Salomons & Pont, 2012; Tang & Wang, 2007; Valente-Pereira, 2014; Weber et al., 2014). Studies have also analyzed the acoustic benefits of parks and green corridors, with models such as the Tranquillity Rating Prediction (TRAP) tool (Pheasant et al., 2010) and the QUIETE index (Brambilla & Gallo, 2016) providing quantitative frameworks.

In this context, this chapter aims to analyze the relationship between urban green spaces, air quality, and noise levels within the study area. Through spatial and statistical analyses, it seeks to uncover how variations in the structure and distribution of green spaces influence environmental parameters. The findings will contribute valuable insights to urban planning, emphasizing the ecological and health benefits of preserving and integrating green spaces for sustainable urban development.

4.2 Database and Methodology

See Fig. 4.1.

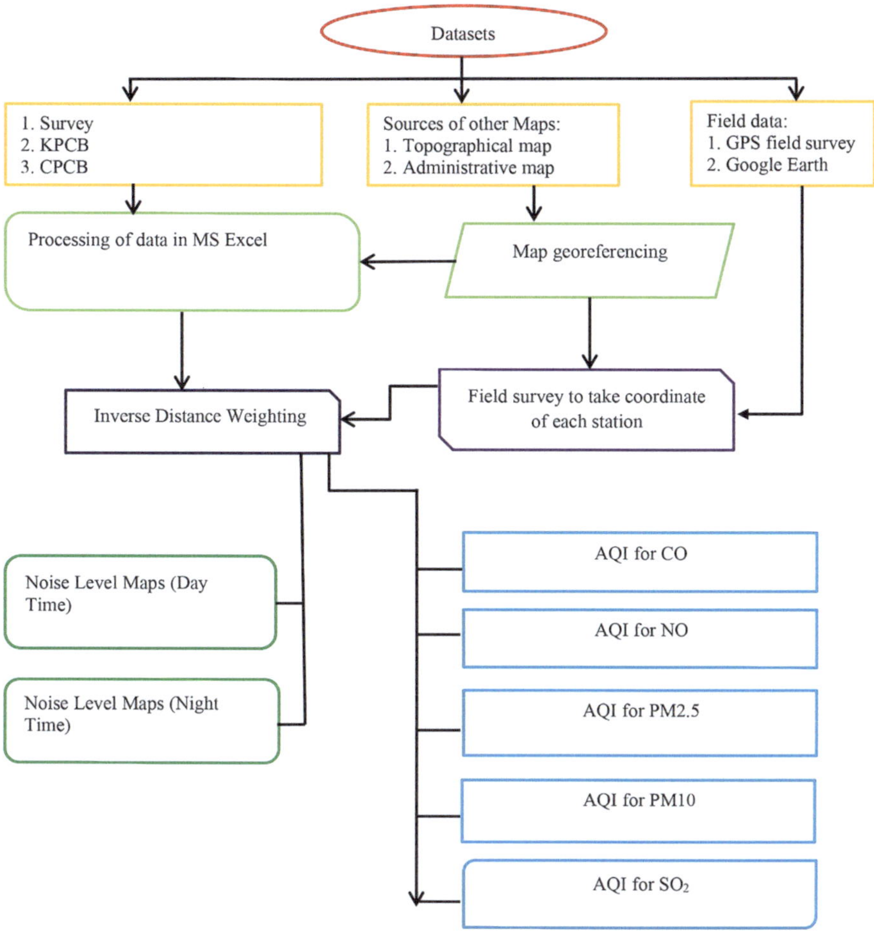

Fig. 4.1 Methodological flowchart of the chapter

4.2.1 Selection of Air Quality Monitoring Stations

A total of seven air quality monitoring stations were selected for this study, namely Ballygunge, Bidhan Nagar, Thakurpukur, Jadavpur, Rabindra Bharati University, Rabindra Sarobar, and Victoria Memorial (Fig. 4.2). These stations were strategically chosen to represent varying environmental conditions within the study area. Some of the selected stations are located in regions characterized by abundant green spaces and a high vegetation index, allowing for the assessment of how dense vegetation influences air quality. Conversely, other stations are situated in zones with low or negligible green cover, where vegetation density is minimal or absent. These include areas dominated by intensive vehicular movement, industrial activities, and high population density, as well as regions susceptible to natural dust or pollen emissions.

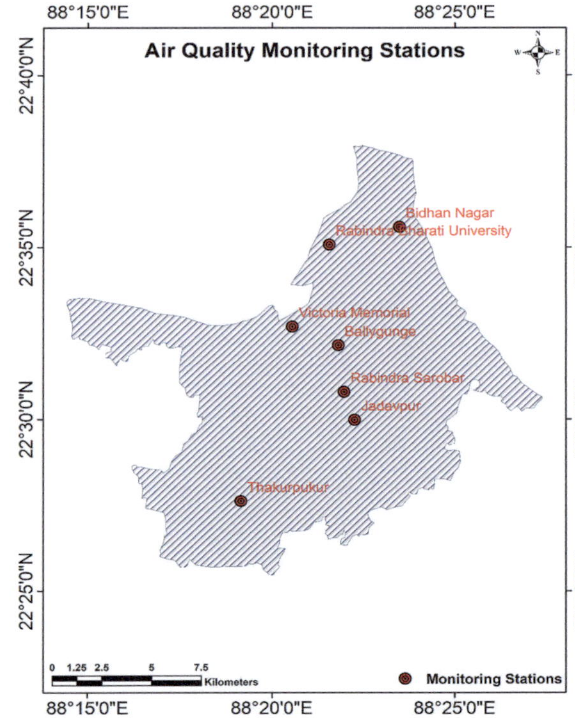

Fig. 4.2 Air quality monitoring stations

The deliberate inclusion of monitoring stations from both green-dominant and green-deficient zones ensures a comprehensive and balanced spatial coverage. This selection approach facilitates comparative analysis between areas of differing ecological characteristics, thereby providing a clearer understanding of the relationship between vegetation cover and air quality variations across the Kolkata Municipal Corporation. Monitoring sites with minimal green cover, positioned in high-traffic corridors and urban cores, offer critical insights into pollution concentrations in heavily urbanized environments, while those with rich vegetation serve as reference points for assessing the mitigating impact of urban green infrastructure on ambient air quality.

4.2.2 Air Quality Monitoring Parameters

The monitoring of key air quality parameters namely carbon monoxide (CO), nitrogen oxide (NO), sulfur dioxide (SO_2), and particulate matter ($PM_{2.5}$ and PM_{10}) is essential for evaluating the extent of air pollution and its implications for both the environment and public health (Fig. 4.1). These parameters are measured using advanced analytical instruments and standardized techniques, ensuring accuracy and consistency across different monitoring sites. Continuous data collection is carried out through a network of fixed air quality monitoring stations, strategically located to provide representative coverage of diverse environmental and land use conditions across the study area. The recorded data are invaluable for a range of purposes, including regulatory compliance, scientific analysis, and public health policy formulation. Moreover, these datasets help identify spatial and temporal pollution trends, trace pollution sources, and assess their correlation with the distribution of urban green spaces. Ultimately, such comprehensive monitoring serves as a critical foundation for developing effective air quality management strategies and promoting sustainable urban environmental planning.

4.2.3 Selection of Noise Level Monitoring Stations

A total of nine noise level monitoring stations were established at strategically selected locations, namely SSKM Hospital, Gole Park, Kolkata HQ (WBPCB), Patauli, New Market, RG Kar Hospital, Tollygunge, Bag Bazar, and Tartala (Fig. 4.3). These monitoring sites were carefully chosen to capture the spatial variation in noise intensity across areas with differing levels of vegetation cover. Some stations are located in green-rich zones characterized by a high vegetation index, enabling the assessment of the influence of greenery on noise attenuation. In contrast, other stations were selected from highly urbanized regions with low or negligible green cover, where noise pollution tends to be more severe due to dense infrastructure, heavy vehicular movement, and the presence of industrial activities. The placement of monitoring stations in these contrasting environments ensures a comprehensive understanding of noise dynamics across the urban landscape. Furthermore, both day time and night time noise levels were recorded at each station to facilitate a comparative analysis of diurnal variations in sound intensity (Fig. 4.1), thereby providing valuable insights into the patterns and sources of urban noise pollution.

4.2.4 Inverse Distance Weighting for Interpolation

Inverse Distance Weighting (IDW) is a commonly applied spatial interpolation method within Geographic Information Systems (GIS), used to estimate unknown values at unsampled locations based on the proximity and values of surrounding sampled points. The underlying principle of IDW is that spatially closer points exert a greater influence on the interpolated value than those located farther away. This method is grounded in the concept of spatial autocorrelation, which assumes that locations situated near one another tend to exhibit similar characteristics or values. Such an assumption makes IDW particularly effective for modeling spatially continuous phenomena, including air pollution concentration, temperature variation, elevation, and noise levels.

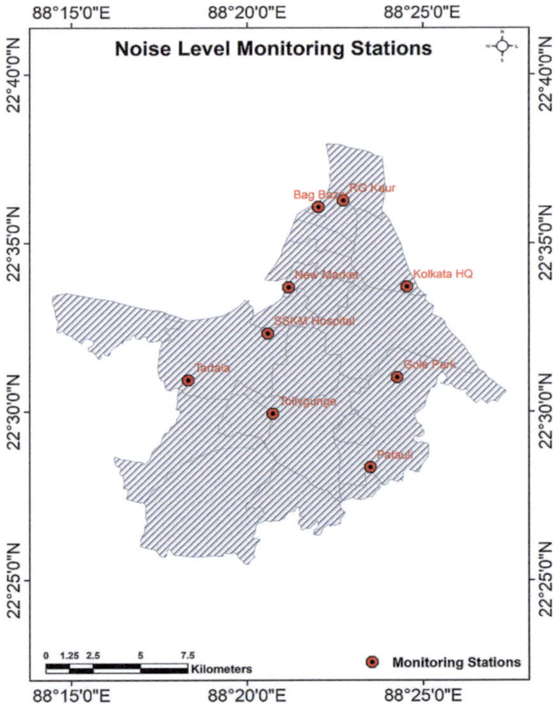

Fig. 4.3 Noise level monitoring stations

In this approach, each sampled point is assigned a weight inversely proportional to its distance from the point being estimated. These weights are typically calculated using a power-law function, where the power parameter controls the rate at which influence decreases with increasing distance. A higher power value causes a rapid decline in influence from distant points, whereas a lower value allows a smoother transition in spatial influence. By integrating these distance-based weights, IDW produces interpolated surfaces that effectively represent the spatial distribution and intensity gradients of environmental variables. Owing to its simplicity, computational efficiency, and accuracy in depicting spatial trends, IDW remains a robust and widely adopted tool for environmental and geographical analyses in GIS-based studies.

Inverse Distance Weighting

$$\mathrm{vp} = \frac{\sum_{i=1}^{n} \frac{1}{d}\mathrm{vi}}{\sum_{i=1}^{n} \frac{1}{d}}$$

4.2.5 Normalized Difference Vegetation Index

The Normalized Difference Vegetation Index (NDVI) is a well-established vegetation indicator extensively utilized in remote sensing and Geographic Information Systems (GIS) to assess the health, density, and spatial distribution of vegetation cover across an area (Fig. 4.4). NDVI is computed from satellite or aerial imagery using reflectance values obtained from the red and near-infrared (NIR) spectral bands. The underlying principle of NDVI is based on the spectral characteristics of vegetation, healthy plants absorb most visible light in the red spectrum for photosynthesis while reflecting a large portion of near-infrared radiation. In contrast, stressed or sparse vegetation reflects more visible light and absorbs less near-infrared radiation.

$$\mathrm{NDVI} = \frac{\text{Near Infrared (NIR)} - \text{Red}}{\text{Near Infrared (NIR)} + \text{Red}}$$

Here, NIR represents the reflectance in the near-infrared band, while *Red* denotes the reflectance in the visible red band. NDVI values typically range from -1 to $+1$, where higher values signify dense and healthy vegetation, whereas lower or negative values indicate barren surfaces, urban areas, or degraded vegetation.

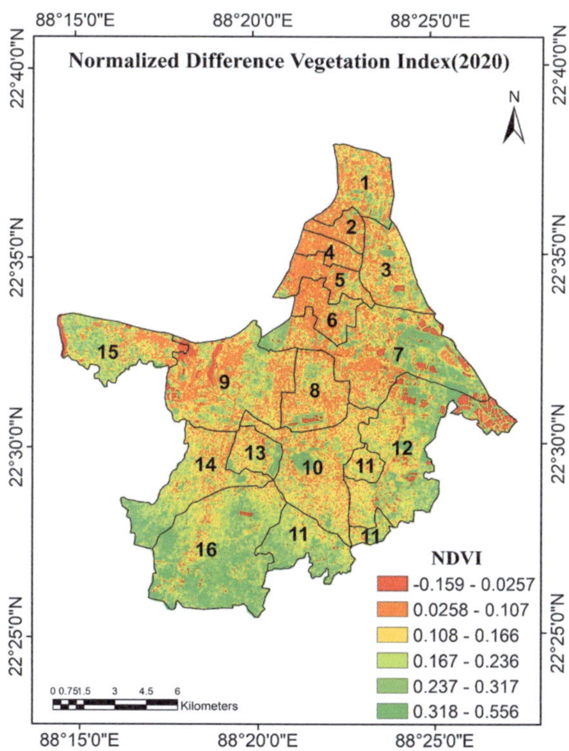

Fig. 4.4 Normalized difference vegetation index (2020)

NDVI serves a wide range of applications across disciplines. In urban environmental studies, it aids in the evaluation of urban green space distribution, vegetation health, and the effectiveness of green infrastructure planning. Ecologists employ it for ecosystem monitoring, while climatologists use it to track vegetation response to climatic variations. Due to its ability to deliver consistent, large-scale, and temporally comparable data, NDVI remains a critical tool for researchers, planners, and policymakers focused on sustainable land management, urban ecological assessment, and environmental change detection.

4.3 Results and Discussion

4.3.1 Concentration Levels of Different Parameters of Air Quality and Their Associations with Green Spaces

Table 4.1 and Fig. 4.5 present a detailed summary of air quality measurements across various monitoring stations in Kolkata, covering major pollutants such as $PM_{2.5}$, PM_{10}, nitrogen oxide (NO), carbon monoxide (CO), and sulfur dioxide (SO_2). These concentrations were recorded over a 24-h period and compared against standard limits prescribed by regulatory authorities.

At Ballygunge, $PM_{2.5}$ (225.42 $\mu g/m^3$) and PM_{10} (290.85 $\mu g/m^3$) levels substantially exceed permissible limits (60 and 100 $\mu g/m^3$, respectively). NO levels are also high at 165.1 $\mu g/m^3$, while CO (2.16 $\mu g/m^3$) and SO_2 (9.91 $\mu g/m^3$) remain within

Table 4.1 Different parameters-based air quality statistics

Stations	Latitude	Longitude	$PM_{2.5}$	PM_{10}	NO	CO	SO_2
			Standard ($\mu g/m^3$) in 24 h				
			60	100	80	4	80
Ballygunge	22.53597222	88.36352568	225.42	290.85	165.1	2.16	9.91
Bidhan Nagar	22.59316762	88.39140858	132.68	237.57	69.23	2.14	27.85
Thakurpukur	22.46026112	88.31915205	87.81	159.16	11.62	0.76	4.75
Jadavpur	22.49974494	88.37103054	26.46	119.01	28	0.89	11.67
Rabindra Bharati University	22.58454804	88.35939401	222.94	524.65	68.27	2.6	11.6
Rabindra Sarobar	22.5132309	88.36619677	124.24	167.32	53.97	1.55	29.13
Victoria Memorial	22.54496186	88.34256622	96.36	178.39	79.96	1.14	7.41

Date of data acquisition—13, 14, 15, 19, 20 January, 2020
Source CPCB (2020)

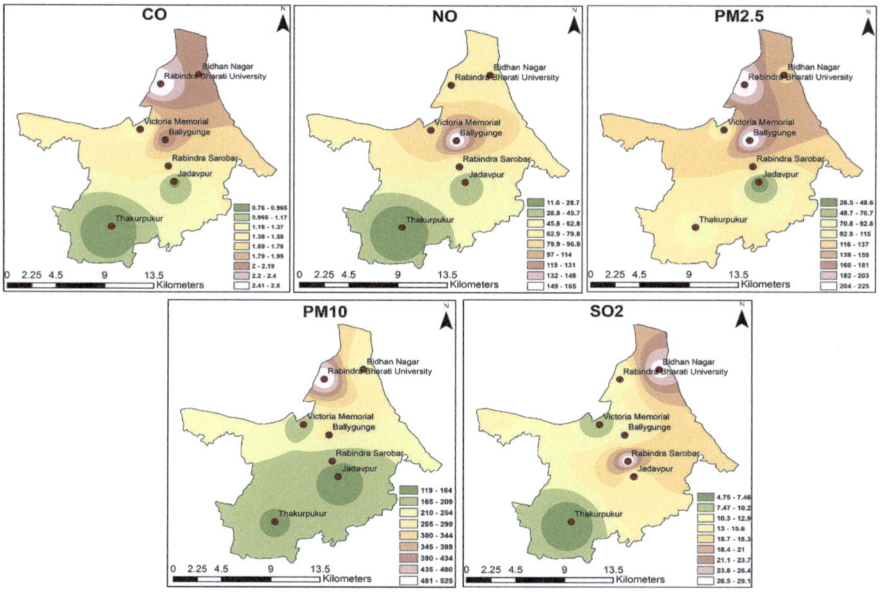

Fig. 4.5 Concentration levels of different parameters of air quality

acceptable limits. The elevated pollutant levels correspond with the low vegetation index and limited green cover in the area. Similarly, Bidhan Nagar, situated in a highly urbanized downtown zone, records $PM_{2.5}$ at 132.68 μg/m^3 and PM_{10} at 237.57 μg/m^3, both above standards. Although NO (69.23 μg/m^3) and CO (2.14 μg/m^3) are within tolerable limits, SO_2 reaches 27.85 μg/m^3, indicating local atmospheric stress due to dense built-up areas and minimal greenery.

In contrast, Thakurpukur demonstrates relatively improved air quality, with $PM_{2.5}$ at 87.81 μg/m^3 and PM_{10} at 159.16 μg/m^3, still above permissible levels but significantly lower than other zones. NO, CO, and SO_2 concentrations remain well within safe limits, attributed to the abundance of vegetation and green cover. Jadavpur, surrounded by large green areas such as Kolkata Golf Club and Kolkata Golf City, records the lowest $PM_{2.5}$ concentration (26.46 μg/m^3), while only PM_{10} (119.01 μg/m^3) slightly exceeds the threshold, reinforcing the mitigating role of greenery in air pollution reduction.

Rabindra Bharati University, a highly urbanized region, exhibits severe air contamination, with $PM_{2.5}$ at 222.94 μg/m^3 and PM_{10} at 524.65 μg/m^3. NO (68.27 μg/m^3) and CO (2.6 μg/m^3) levels are also elevated, reflecting the impact of dense construction and vehicular emissions in areas with negligible green spaces. Rabindra Sarobar shows moderate pollution, with $PM_{2.5}$ (124.24 μg/m^3) and PM_{10} (167.32 μg/m^3) above limits, while SO_2 (29.13 μg/m^3) remains a concern due to declining vegetation density.

At Victoria Memorial, pollutant levels are comparatively lower—$PM_{2.5}$ (96.36 μg/m^3) and PM_{10} (178.39 μg/m^3) owing to extensive greenery and open landscapes.

NO (79.96 μg/m^3), CO (1.14 μg/m^3), and SO_2 (7.41 μg/m^3) all fall within acceptable limits, emphasizing the positive environmental role of urban green zones.

Overall, the data clearly demonstrate that areas rich in vegetation and green spaces exhibit lower pollutant concentrations, while highly urbanized and concretized regions show severe air quality degradation. Trees and vegetation act as natural air purifiers, filtering particulates, absorbing gaseous pollutants, and moderating urban temperatures. Hence, the integration of green infrastructure in urban landscapes is critical for improving air quality, reducing health risks, and fostering sustainable urban living.

4.3.2 Concentration of Noise Levels and Its Associations with Green Spaces

Noise pollution represents a significant urban environmental challenge, posing serious risks to human health and quality of life. Studies have estimated that the economic costs of noise pollution range from 0.2 to 2% of the European Union's GDP (Bolund & Hunhammar, 1999). In densely populated metropolitan regions, urban green spaces play a vital role in mitigating noise intensity, with their effectiveness largely dependent on their extent, vegetation density, and proximity to noise sources. Within this context, green spaces are recognized as an essential component of the urban ecosystem, contributing to the preservation of ecological balance and promoting environmental sustainability in cities.

Tables 4.2 and 4.3, along with Figs. 4.6 and 4.7, present a comprehensive dataset of day time and night time noise levels, expressed in Leq dB(A), recorded at various monitoring stations across the Kolkata Municipal Corporation (KMC) throughout the year.

At SSKM Hospital, noise levels remain relatively stable, fluctuating between 43 dB(A) in September and 52 dB(A) in January. The consistently low readings can be attributed to the hospital's proximity to extensive green spaces such as the Maidan, which serves as a major vegetated buffer zone. Gole Park, on the other hand, shows substantial month-to-month variation, with noise levels peaking at 70 dB(A) in May and declining thereafter an indication of limited vegetation cover and possibly seasonal or event-driven disturbances.

The Kolkata HQ (WBPCB) station maintains a relatively steady acoustic profile, with minor fluctuations throughout the year. Being located in a densely built-up downtown area, the lack of green spaces results in persistently high noise exposure. In contrast, Patauli demonstrates significant variability, with noticeable peaks in April and June (62 dB(A)), while maintaining lower values during months with higher green cover activity, reflecting the noise-absorbing capacity of vegetation in its vicinity.

New Market, situated in a commercial hub, records consistently high noise levels between 60 and 69 dB(A) due to heavy traffic and dense construction, coupled with

Table 4.2 Noise level data (day time) in Leq dB(A)

Station location	Latitude	Longitude	January	February	March	April	May	June	July	August	September	October	November	December
SSKM Hospital	22.53877	88.34314	52	49	51	50	49	51	48	44	43	44	52	48
Gole Park	22.51700	88.40439	59	62	64	67	70	65	67	64	67	67	60	60
Kolkata HQ (WBPCB)	22.56185	88.40902	65	64	61	64	63	66	65	64	64	65	63	63
Patauli	22.47252	88.39159	50	44	44	62	47	62	62	62	59	50	61	67
New Market	22.56150	88.35289	66	65	61	60	61	61	62	61	60	61	63	69
RG Kar Hospital	22.60449	88.37867	56	52	45	44	45	44	45	44	43	44	44	44
Tollygunge	22.49902	88.34550	66	66	66	66	64	65	66	65	66	66	66	66
Bag Bazar	22.60128	88.36695	78	79	78	77	65	69	67	63	54	65	74	70
Tartala	22.51556	88.30533	65	65	64	65	66	66	65	66	65	64	64	65

Source CPCB (2020)

Table 4.3 Noise level data (night time) in Leq dB(A)

Station location	Latitude	Longitude	January	February	March	April	May	June	July	August	September	October	November	December
SSKM Hospital	22.53877	88.34314	51	49	53	52	50	53	50	46	45	46	47	44
Gole Park	22.51700	88.40439	55	55	60	65	70	65	68	64	68	67	61	60
Kolkata HQ (WBPCB)	22.56185	88.40902	62	64	59	58	60	64	60	57	58	60	56	57
Patauli	22.47252	88.39159	51	44	48	63	49	58	61	64	59	50	61	69
New Market	22.56150	88.35289	67	65	61	60	62	61	62	62	60	61	60	62
RG Kar Hospital	22.60449	88.37867	57	52	45	44	45	44	45	43	43	44	44	43
Tollygunge	22.49902	88.34550	61	62	62	62	60	60	61	59	60	62	61	62
Bag Bazar	22.60128	88.36695	78	78	77	75	65	68	66	63	53	61	72	69
Tartala	22.51556	88.30533	61	61	61	62	61	62	62	61	62	60	60	60

Source CPCB (2020)

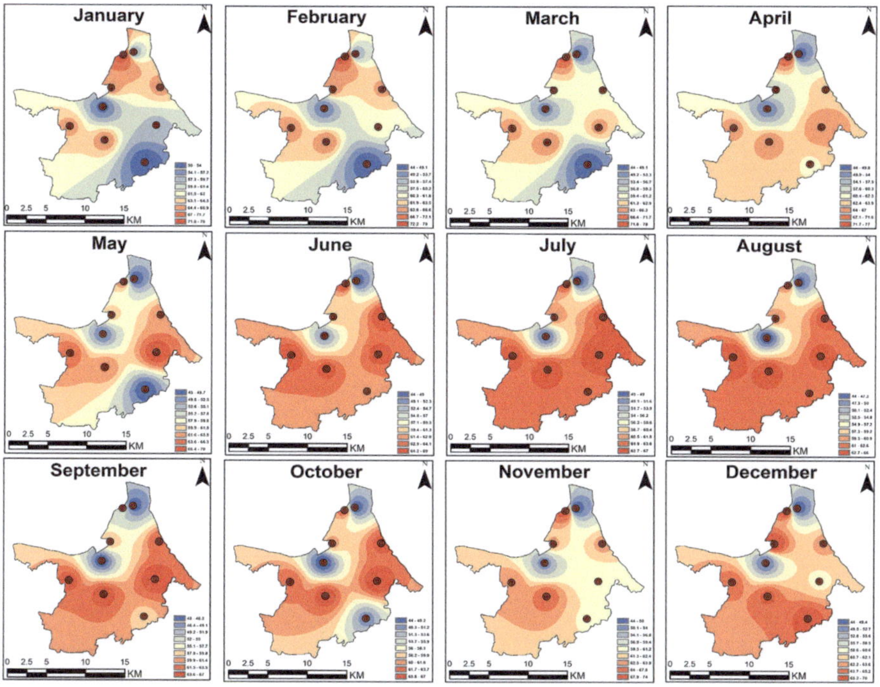

Fig. 4.6 Day time noise level

minimal greenery. Meanwhile, RG Kar Hospital exhibits lower and stable readings ranging between 43 and 56 dB(A), primarily because it lies within a designated silence zone where noise regulations are enforced.

Similarly, Tollygunge records consistently high noise levels averaging 66 dB(A) throughout the year, indicating limited vegetation cover and constant exposure to urban activities. Bag Bazar and Tartala display marked fluctuations, with Bag Bazar peaking at 79 dB(A) in February before declining later in the year, suggesting the influence of seasonal or anthropogenic factors. Both stations are situated in highly concretized zones with negligible greenery, exacerbating sound reflection and amplification.

Overall, the analysis highlights clear spatial contrasts between green-rich and concrete-dominated zones. Stations located in or near green spaces consistently recorded lower noise levels, whereas areas with minimal vegetation reported elevated and fluctuating readings. These findings underscore the indispensable role of urban green infrastructure in noise attenuation, supporting the need for integrating vegetation buffers, parks, and tree belts into urban planning and environmental management frameworks to enhance the acoustic comfort and well-being of city residents.

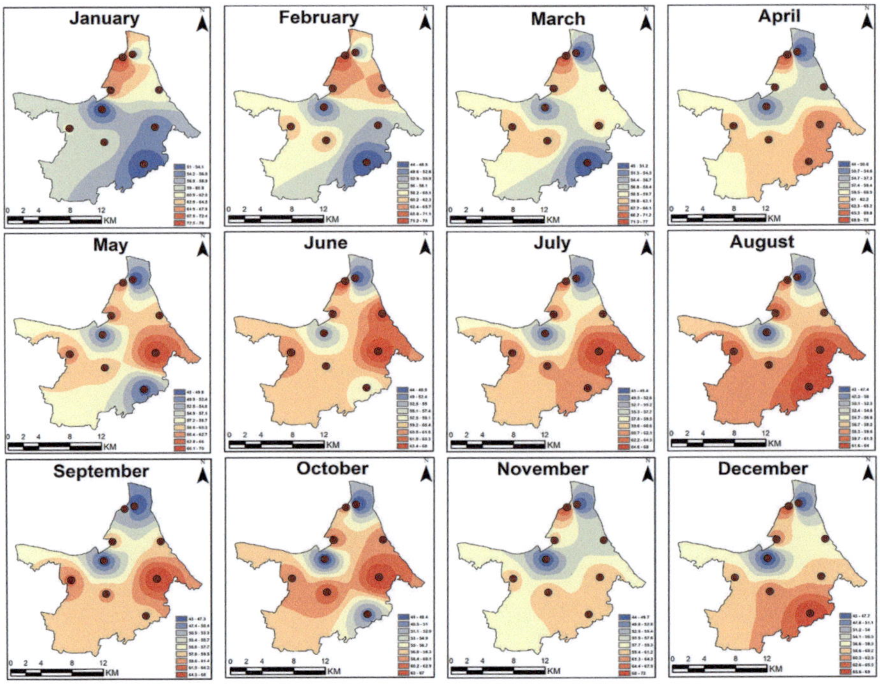

Fig. 4.7 Night time noise level

Chapter 5
Impacts of Urban Green Spaces on Human Health and Well-Being

5.1 Background

Urban green spaces play a crucial role in enhancing both psychological and physical well-being. Numerous studies have shown that regular exposure to natural environments helps reduce stress, anxiety, and depression, thereby improving mental health and emotional stability. These spaces also foster social interaction and community cohesion by providing environments for recreation, leisure, and relaxation. From a physical health perspective, urban greenery encourages outdoor physical activities, which help reduce obesity rates, enhance cardiovascular function, and contribute to faster recovery from illness and lower mortality rates (Hartig et al., 2015; WHO, 2016; Huang et al., 2017). Even brief exposure to green environments such as walking in a park or viewing greenery from a window has been found to enhance well-being, accelerate post-operative recovery (Ulrich, 1984), improve heart health (Paquet et al., 2013), boost mental well-being (Sturm & Cohen, 2014), and contribute to better self-reported health and longevity (Maas et al., 2006; Mitchell & Popham, 2008).

Urban green spaces exist in a hierarchical structure, ranging from small neighborhood parks and playgrounds to extensive city parks and national reserves. Neighborhood parks commonly integrate recreational amenities, sports facilities, and passive green zones (Ersoy, 2015). For these areas to be effective, they must be accessible, safe, aesthetically designed, and inclusive. Key design factors influencing their utility include walkability, cycling pathways, lighting, tree cover, water features, and overall green area size (Kaczynski et al., 2008; Schipperijn et al., 2013).

A growing body of literature highlights the positive link between green space availability and urban quality of life (Martinez-Gonzalez et al., 2001; Cicea & Pîrlogea, 2011; Maas et al., 2009). Access to visually appealing and well-maintained green areas is increasingly recognized as a vital component of a healthy urban lifestyle, as it alleviates stress and contributes to emotional balance. The World Health Organization (WHO) emphasizes that green spaces improve physical activity, air quality, social well-being, and noise reduction, while acknowledging possible minor risks

© The Author(s), under exclusive license to Springer Nature Switzerland AG 2026
M. B. Ali and S. Jamal, *Urban Green Spaces and Environmental Sustainability*,
https://doi.org/10.1007/978-3-032-15322-7_5

related to increased recreational activity (WHO, 2010d). The WHO's subsequent reports advocate for publicly accessible parks and open spaces equipped for diverse age groups to encourage active living and reduce health inequalities (WHO, 2013).

The ongoing decline in urban greenery has heightened global concern regarding its implications for public health (Cicea & Pîrlogea, 2011). Physical inactivity identified by the WHO as one of the leading global health risks affects up to 80% of individuals in certain regions. Regular physical activity is known to improve heart health, reduce risks of diabetes and cancer, and enhance mental well-being. In line with this, the WHO's Global Action Plan on Physical Activity (2018–2030) promotes the vision of "More Active People for a Healthier World," emphasizing the expansion of green, recreational, and sports spaces to foster active, healthy lifestyles (WHO, 2018).

Recent scholarship continues to affirm the strong connection between urban green spaces and mental wellness, providing growing evidence of their role in psychological restoration and stress reduction (Sarkar et al., 2018; Callaghan et al., 2021). Kaplan and Kaplan (1989) introduced the concept of psychological restoration, suggesting that natural settings counteract mental fatigue induced by urban stress. Empirical studies further demonstrate that proximity to green spaces can lower cortisol levels (Roe et al., 2013), buffer the impact of stressful life events (Van den Berg et al., 2010), alleviate maternal depression (McEachan et al., 2016), promote social cohesion (Gonzalez & Kirkevold, 2016), and improve overall psychological well-being (Annerstedt et al., 2012; Callaghan et al., 2021).

Despite extensive research on the environmental functions of urban greenery, its direct health implications have not been sufficiently explored. Nevertheless, the multifaceted health benefits of green spaces align with the WHO's recommendations regarding accessibility and public health promotion (WHO, 2016). Accordingly, this chapter seeks to examine the interrelationships between urban green spaces and the psychosocial and physical health of residents in the study area, drawing upon field survey-based primary data to support the analysis.

5.2 Database and Methodology

5.2.1 Interview Cum Schedule Method

The interview-cum-schedule method is a widely adopted data collection technique, particularly useful in social science research. It integrates the strengths of both structured interviews and questionnaires, offering a systematic yet flexible means of gathering primary data from selected respondents. In this approach, trained enumerators or interviewers personally administer a pre-designed schedule containing a set of standardized questions to participants. This ensures uniformity in the responses collected while allowing the interviewer to clarify questions or probe for additional details when necessary.

The method effectively balances the rigidity of structured surveys with the interactive nature of interviews, thus improving the accuracy and reliability of the collected information. Since all respondents are asked the same set of questions in a consistent sequence, this approach minimizes interviewer bias and enhances the comparability of data across different individuals or groups. Furthermore, it allows researchers to collect both quantitative and qualitative insights, capturing not only factual data but also personal perceptions and experiences.

One of the key advantages of the interview-cum-schedule method is its ability to reach diverse segments of the population, including individuals with varying levels of literacy who might otherwise struggle with self-administered questionnaires. It also allows the interviewer to ensure completeness of responses and maintain control over the data collection process, leading to higher response accuracy.

Overall, the interview-cum-schedule method serves as a comprehensive and efficient data collection tool, providing a structured framework for gathering rich, reliable, and comparable data while maintaining a personal and responsive approach essential for field-based studies.

5.2.2 Sampling

Random sampling is one of the most effective and widely applied techniques in research methodology, ensuring that each individual within a population has an equal probability of selection. This statistical approach enhances the representativeness of the sample by capturing the diversity of the population, thereby minimizing selection bias and improving the overall reliability of findings. It enables researchers to make valid generalizations about the entire population based on data obtained from a randomly selected subset, thus providing a more accurate and unbiased reflection of population characteristics. In the present study, the Kolkata Municipal Corporation (KMC) area, which comprises 16 Boroughs (Fig. 5.1), served as the sampling framework. From each borough, 30 respondents were selected through the random sampling technique, with the selection process guided by their proximity to urban green spaces. Consequently, the total sample size amounted to 480 respondents, representing a balanced and diverse cross-section of the KMC population.

5.2.3 Likert Scale

The Likert Scale, developed by Rensis Likert, is a widely recognized psychometric instrument used to measure individuals' attitudes, perceptions, and opinions toward a given subject. It provides a systematic framework that enables respondents to express the intensity of their agreement or disagreement with a set of statements related to a specific issue. Typically, the scale includes a sequence of statements, and respondents rate each one on an ordinal scale ranging from "Strongly Disagree" to

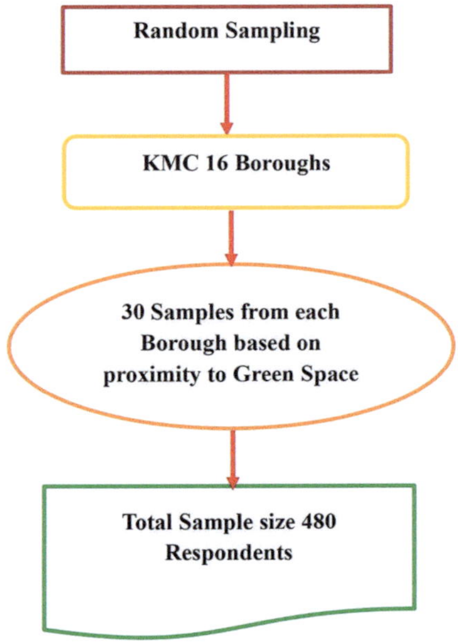

Fig. 5.1 Methodological flowchart of the chapter

"Strongly Agree" with intermediate categories such as "Disagree," "Neutral," and "Agree." This method captures the direction and relative strength of respondents' attitudes, allowing for a nuanced understanding of subjective viewpoints.

The Likert Scale is particularly effective in assessing qualitative perceptions where numerical precision is secondary to the ordering of responses. It generates ordinal data, which reflects the relative ranking of opinions but not the exact magnitude of difference between them. In this chapter, the researcher employed a five-point Likert Scale to evaluate respondents' perceptions of the psychological impacts of urban green spaces, focusing on five key psychological parameters. This approach facilitated the quantification of subjective attitudes toward the influence of green spaces on mental well-being in the study area.

5.2.4 Karl Pearson's Coefficient of Correlation

Karl Pearson's coefficient of correlation, symbolized as "r", is a fundamental statistical tool used to measure the strength and direction of a linear relationship between two continuous variables. It evaluates how closely the paired data points fit along a straight line, thereby indicating the degree of association between them. A positive correlation ($r > 0$) signifies that as one variable increases, the other tends to increase correspondingly, whereas a negative correlation ($r < 0$) indicates that an increase in one variable is typically accompanied by a decrease in the other.

$$\textbf{\textit{Karl Pearson's Coefficient of correlation}}$$
$$r = \frac{n\Sigma x_i y_i - \Sigma x_i y_i}{\sqrt{n\Sigma x_i^2 - (\Sigma x_i)^2} \times \sqrt{n\Sigma y_i^2 - (\Sigma y_i)^2}}$$

This coefficient provides a quantitative measure of the linear relationship between two variables, with values ranging from − 1 to + 1, where + 1 indicates a perfect positive correlation, − 1 represents a perfect negative correlation, and 0 denotes no correlation.

In this chapter, Karl Pearson's coefficient of correlation has been applied to determine the association between urban green spaces and the psycho-physical health of respondents (Tok et al., 2020). This statistical technique enables the researcher to interpret how variations in the availability or quality of urban green spaces influence the mental and physical well-being of individuals, offering valuable insights into the broader implications of urban environmental quality on human health.

5.2.5 Paired Two Tailed t-Test

The paired two-tailed t-test is a statistical technique employed to compare the means of two related datasets in order to determine whether a significant difference exists between them. It is termed "two-tailed" because it evaluates the possibility of a difference in either direction assessing whether one condition is significantly higher or lower than the other. The test operates under two hypotheses: the null hypothesis (H_0), which assumes no significant difference between the paired observations, and the alternative hypothesis (H_1), which posits the presence of a significant difference. To ensure the validity of the results, certain assumptions must be satisfied, including the normal distribution of differences and the random selection of samples.

This method is particularly effective for analyzing before-and-after scenarios or matched pairs, making it a valuable statistical tool across disciplines such as environmental studies, medicine, psychology, and social sciences. In the context of this chapter, the paired two-tailed t-test has been applied to evaluate whether a statistically significant association exists among the studied variables, specifically examining the relationship between urban green spaces and the psycho-physical health of respondents. This approach enables a rigorous testing of hypotheses, ensuring that the observed relationships are not due to random variation but reflect genuine patterns within the data.

5.2.6 Hypotheses

The present study shall make an attempt to test the hypothesis that,

- ***There is significant effect of availability of green space on resident's psycho-social health.***
- ***There is significant effect of availability of green spaces on physical health.***

5.3 Results and Discussion

This chapter is primarily founded on primary data collected through the random sampling technique, with respondents' views gathered using the interview-cum-schedule method to analyze the interrelationship between urban green spaces and psycho-social and physical health. The study area, Kolkata Municipal Corporation (KMC), encompasses 16 boroughs covering a total area of 206 km^2. From each borough, 30 respondents were selected for interviews, 15 from areas with high concentrations of green spaces and 15 from areas with very low green space availability ensuring balanced representation across contrasting environmental conditions. The data collected from respondents were systematically tabulated and analyzed to test the two hypotheses established in the methodology section.

Table 5.1 and Fig. 5.2 display the distribution of responses based on the five-point Likert Scale, designed to measure psychological well-being in relation to green space exposure across five psyche-related parameters. The results reveal that 34% of respondents strongly agree, and 45% agree that they feel stress-free and emotionally uplifted when in or around green spaces. About 14% remain neutral, while only 7% either disagree or strongly disagree. These findings highlight that a substantial majority of participants associate urban green spaces with reduced stress, enhanced happiness, improved mood, and emotional stability. The analysis underscores the positive psychological influence of green environments, affirming their critical role in fostering mental well-being and emotional resilience among city residents.

Hypothesis 1

H_0 ***There is no significant effect of availability of green space on resident's psycho-social health.***

H_a ***There is significant effect of availability of green space on resident's psycho-social health.***

Table 5.1 Responses of respondents on Likert's scale

Questions	Scale				
Do you feel stress free when you are in or around of a green space?	Strongly agree	Agree	Undecided	Disagree	Strongly disagree
Do you feel happy when you are in a green space?					
Do you feel that when you are in and around of a green space then your mood is joyful?					
When you are in depression and anxiety then do you go towards green space for feeling better?					
Do you visit green space when you are emotionally unstable?					
Total responses	164	216	68	24	8
Total responses (%)	34	45	14	5	2

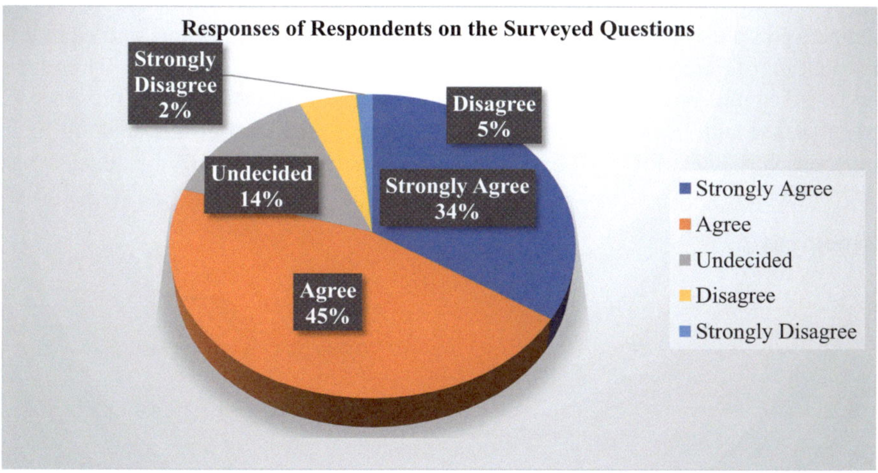

Fig. 5.2 Responses of respondents on Likert's scale

Table 5.2 Correlation statistics of the tested hypothesis

Variables	Correlation (r)	P value
Availability of green space	0.675	0.005**
Psycho-social health		

** Correlation is significant at the 0.01 level (two-tailed)

The results presented in Table 5.2 illustrate the statistical relationship between two primary variables availability of green space and psycho-social health—through correlation analysis. The correlation coefficient (r) measures the direction and strength of the linear association between these variables, ranging from -1 to $+1$. Here, the correlation coefficient value of 0.675 indicates a strong positive relationship between the availability of green spaces and psycho-social health. This suggests that individuals who live in neighborhoods with greater accessibility to green areas tend to report higher levels of psychological well-being, reduced stress, and improved social interactions.

The p-value, which determines the statistical significance of the correlation, is 0.005, marked by a double asterisk to denote a high level of significance at the 1% threshold (two-tailed test). This small p-value indicates that the likelihood of this correlation occurring by chance is extremely low, thus confirming that the observed relationship between green space availability and psycho-social health is statistically significant and reliable.

These findings have important implications for urban development and public health policy. The strong positive correlation highlights the critical role of urban green spaces in enhancing psychological well-being, promoting social cohesion, and fostering emotional stability among residents. It reinforces the idea that urban planners and policymakers should prioritize preservation, restoration, and equitable distribution of green spaces to improve the overall quality of urban life. Therefore, the tested hypothesis provides compelling empirical evidence that the presence of green spaces significantly contributes to better psycho-social health, making them indispensable assets for creating sustainable, livable, and mentally healthier urban environments.

Hypothesis 2

H_0 ***There is no significant effect of availability of green spaces on physical health.***

H_a ***There is significant effect of availability of green spaces on physical health.***

The results presented in Table 5.4 depict the statistical relationship between availability of green spaces and physical health using correlation analysis. The correlation coefficient (r) of 0.4 indicates a moderate positive relationship, implying that an increase in the availability of green spaces is generally associated with an improvement in physical health. This means that individuals residing in areas with greater green space coverage tend to experience better physical well-being. However, the moderate strength of the correlation suggests that while green spaces positively influence health, other factors such as lifestyle choices, nutrition, and access to healthcare may also play a significant role in determining overall physical health outcomes.

The p-value of 0.02, marked with a double asterisk (**), denotes a statistically significant relationship at the 5% level (two-tailed test). This low p-value indicates that the observed correlation is unlikely to have occurred by random chance, affirming that the association between green space availability and physical health is meaningful and reliable.

The implications of these findings are particularly relevant for public health planning and urban policy. The presence of green spaces provides residents with opportunities for physical activity, such as walking, jogging, or cycling, which contribute to improved cardiovascular health, reduced obesity, and lower stress levels (Table 5.3). Moreover, green environments can enhance air quality and promote outdoor recreation, further supporting healthy lifestyles.

In summary, this hypothesis confirms a moderate but statistically significant positive correlation between the availability of green spaces and physical health. While green spaces evidently foster better health outcomes, the findings also suggest the need for an integrated approach that combines environmental planning with broader health and wellness initiatives to maximize their benefits.

5.3.1 Correlation Between Distance from Green Spaces; and Frequency and Duration of Their Utilization

An examination of Table 5.5 reveals the statistical correlations between the variable "Distance from Green Spaces" (X_1) and two key aspects of physical activity "Frequency of Physical Activities" (Y_1) and "Time Duration of Physical Activities" (Y_2). The computed correlation coefficients (r) demonstrate negative relationships in both cases, indicating that as the distance from green spaces increases, both the frequency and duration of physical activities tend to decrease. Specifically, the correlation between distance and frequency of physical activities is − 0.219, while the correlation between distance and time duration of physical activities is − 0.351. These negative coefficients suggest that individuals who live closer to green spaces are more likely to engage in physical activities and to sustain them for longer durations than those residing farther away.

The statistical significance at the 0.01 level (two-tailed) confirms that these correlations are unlikely to be the result of random variation, implying a meaningful

Table 5.3 Responses recorded during field survey

Questions							
Distance between residence and green space	0–0.5 KM	0.6–1.0 KM	1.1–1.5 KM	1.6–2.0 KM	2.1–2.5 KM	2.6–3.0 KM	> 3.1 KM
	32	17	12	16	11	8	4
Mode of reaching green space	Walking	Vehicle	Cycling				
	47	32	21				
Frequency of use of green space for physical activities	Once per week	Twice per week	Trice per week	Four times in a week	Five times in a week	Six times in a week	Every day in a week
	52	20	8	7	2	6	5
Time spend in green space	20 min and less	20–40 min	40–80 min	80 min and above			
	21	52	21	6			
Body mass index	Under weight	Normal weight	Over weight	Obesity			
	12	39	45	4			
Diseases related to respiratory system	Acute lower respiratory infection	Acute upper respiratory infection	Chronic pulmonary disease	Asthma	Pulmonary tuberculosis	Blood pressure	
	2	7	9	7	4	6	
Cardiovascular diseases (CVDs)	Coronary heart disease	Peripheral arterial disease	Rheumatic heart disease	Congenital heart disease	Deep vein thrombosis and pulmonary embolism	Other diseases	
	13	9	10	2	5		

Table 5.4 Correlation statistics of the tested hypothesis

Variables	Correlation (r)	P value
Availability of green spaces	0.4	0.02**
Physical health		

** Correlation is significant at the 0.05 level (2-tailed)

Table 5.5 Correlation statistics of the tested hypothesis

Variables	Distance from green spaces (X_1)
Frequency of Physical Activities (Y_1)	– 0.219*
Time Duration of Physical Activities (Y_2)	– 0.351*

* Correlation is significant at the 0.01 level (two-tailed)

and consistent pattern between accessibility to green spaces and physical activity behavior. In essence, proximity to green spaces serves as a strong motivator for individuals to participate in regular physical exercise.

From a broader perspective, these findings hold important implications for urban planning and public health policy. They emphasize that enhancing accessibility to green spaces can significantly contribute to promoting active lifestyles, improving community health, and fostering overall well-being. By strategically integrating green spaces within residential areas, urban planners can help reduce physical inactivity and related health risks. In summary, Table 5.5 provides clear evidence of significant negative correlations between distance from green spaces and both the frequency and duration of physical activities, underlining the pivotal role of spatial proximity to green environments in encouraging sustained physical engagement and supporting healthier urban living.

5.3.2 *Significance of Green Spaces on Observed Diseases During Field Survey*

Urban green spaces have a substantial influence on public health, particularly in mitigating diseases related to the respiratory and cardiovascular systems. These green environments comprising parks, gardens, and tree-covered areas serve as vital natural assets that improve air quality, encourage physical activity, and enhance overall well-being. Functioning as natural air filters, green spaces absorb harmful pollutants such as carbon dioxide, nitrogen oxides, and particulate matter, while simultaneously releasing oxygen. This process significantly reduces exposure to airborne pollutants, thereby decreasing the incidence of respiratory illnesses like asthma, chronic obstructive pulmonary disease (COPD), acute respiratory infections, and tuberculosis. In addition, green spaces promote regular physical exercise activities such as walking, cycling, or jogging which play a key role in maintaining cardiovascular health by lowering risks associated with conditions like coronary heart disease, hypertension, and peripheral arterial disease. Beyond their physiological benefits, these environments also foster mental relaxation and stress reduction, both of which are known to have positive effects on cardiovascular functioning by reducing stress-induced blood pressure spikes and heart strain. Thus, urban green spaces function as essential preventive healthcare infrastructures that support respiratory health through cleaner air and bolster cardiovascular wellness through enhanced physical and psychological resilience.

Field survey findings revealed notable instances of respiratory and cardiovascular disorders among respondents, with clear associations linked to the availability of green spaces. Among respiratory conditions, acute lower respiratory infections were observed in 2% of respondents, acute upper respiratory infections in 7%, chronic pulmonary diseases in 9%, asthma in 7%, and pulmonary tuberculosis in 4% (Table 5.6). Regarding cardiovascular ailments, coronary heart disease affected

Table 5.6 Symptomatic/clinical diseases observed during field survey

Diseases related to respiratory system	*Acute lower respiratory infection*	*Acute upper respiratory infection*	*Chronic pulmonary disease*	*Asthma*	*Pulmonary tuberculosis*	*Blood pressure*
% of respondents	2	7	9	7	4	6
Cardiovascular Diseases (CVDs)	*Coronary heart disease*	*Peripheral arterial disease*	*Rheumatic heart disease*	*Deep vein thrombosis and pulmonary embolism*	*Other diseases (cholesterol, obesity, etc.)*	
% of respondents	13	9	10	2	5	

13% of participants, peripheral arterial disease 9%, rheumatic heart disease 10%, deep vein thrombosis and pulmonary embolism 2%, while other related conditions such as cholesterol and obesity were reported by 5% (Table 5.6). These results indicate that the prevalence of respiratory and cardiovascular diseases tends to be higher in areas with limited or poorly maintained green spaces. The data underscore the critical importance of integrating and preserving urban greenery as a preventive public health measure, capable of reducing disease prevalence, improving air quality, and enhancing the overall well-being of urban populations.

Chapter 6
Urban Green Space Vulnerability and Future Sustainability Strategies

6.1 Background

Urbanization, a defining global phenomenon, has transformed landscapes and restructured human settlements worldwide. The rapid influx of populations into urban centers has placed immense strain on city infrastructures, particularly on urban green spaces, which are essential for maintaining ecological balance and urban livability (Zhang & Shunfeng, 2003; Sato & Yamamoto, 2005; Roy, 2009; Chen et al., 2011; Liu et al., 2014; Naikoo et al., 2022). Accelerated urban growth has brought about significant environmental, social, and economic challenges, with the vulnerability of green spaces emerging as a key concern. This vulnerability reflects the susceptibility of these vital ecological and recreational assets to multiple stressors in fast-urbanizing environments (Shahfahad et al., 2023). To accommodate growing urban populations, extensive expansion and infrastructure development have been pursued, frequently at the cost of green cover (Byomkesh et al., 2012; Zhou et al., 2012; Luy et al., 2012; Jia, 2013). Such transformations have often led to the conversion of vegetated land into impermeable surfaces, a trend observed across global cities, yielding adverse environmental consequences (Kong & Nakagoshi, 2006; Uy & Nakagoshi, 2007; Liu et al., 2014).

In urban contexts, green spaces perform indispensable ecological and social functions, they mitigate air pollution, generate oxygen, regulate microclimates, support biodiversity, and manage stormwater, while simultaneously contributing to human well-being and urban aesthetics (Rafiee et al., 2009; Singh et al., 2010; Wu, 2014; Kabisch et al., 2016; Xu et al., 2018; Aklibasinda & Ozdarici, 2019; Shahfahad et al., 2021). Recognizing their value for sustainability and quality of life, researchers have increasingly focused on assessing the status, distribution, and vulnerability of urban green spaces (Doygun & Iter, 2007; Hofmann et al., 2011; Badiu et al., 2016; Santos et al., 2016; Sun et al., 2019). As cities expand, land use dynamics shift dramatically (Naikoo et al., 2023), often leading to substantial alterations in ecosystem structure, function, and biodiversity (Fan et al., 2018).

© The Author(s), under exclusive license to Springer Nature Switzerland AG 2026

M. B. Ali and S. Jamal, *Urban Green Spaces and Environmental Sustainability*,
https://doi.org/10.1007/978-3-032-15322-7_6

The vulnerability of urban green spaces stems from a combination of environmental and anthropogenic pressures. Rapid land use changes driven by urban sprawl, infrastructure growth, and industrialization frequently result in diminished green coverage and disrupted ecological systems (Jamal & Ahmad, 2020; Ganaie et al., 2021). Pollution from vehicles and industries further threatens vegetation health and ecosystem resilience (Jamal et al., 2022). Climate change amplifies these threats by altering species composition, increasing temperature stress, and affecting local microclimates. Several studies have explored how these factors land use transformation, pollution, climate variability, and socio-economic inequalities contribute to the degradation and loss of urban green areas (Bardhan et al., 2016). Therefore, addressing the vulnerability of green spaces has become an essential component of sustainable urban planning and environmental management (Lee et al., 2015; McConnells & Walls, 2005).

The advent of remote sensing and Geographic Information Systems (GIS) has revolutionized the assessment of urban green spaces by enabling large-scale, precise, and temporal analyses. Early works, such as Forster (1983), demonstrated the value of Landsat imagery in assessing urban quality and green space distribution (Sudha & Ravindranath, 2000; Sivaramakrishnan et al., 2005; Gupta et al., 2012; Bardhan et al., 2016). Recent studies employ advanced GIS-based analytical techniques, including multi-criteria mapping, spatial weighted overlay, and spatial regression, to evaluate spatial variations in green space vulnerability (Ali & Ahmad, 2018; Shahfahad et al., 2022; Ajmal et al., 2022).

Within the Kolkata Municipal Corporation (KMC) area, effective management and conservation of urban green spaces are fundamental to achieving sustainability (Jamal et al., 2022). In this densely populated and rapidly urbanizing metropolis, these areas function as ecological stabilizers, air purifiers, and recreational zones. Their preservation enhances biodiversity, mitigates the urban heat island effect, and improves public health and overall urban resilience. Therefore, integrating green infrastructure into urban planning is essential for ensuring environmental sustainability and enhancing the quality of life for Kolkata's residents.

Given the mounting pressures of urbanization, evaluating the vulnerability of urban green spaces has become a pressing necessity. This chapter aims to critically assess the methodologies used in evaluating such vulnerabilities, identify the key environmental and socio-economic stressors contributing to them, and project the future development and sustainability of urban green spaces within the study area over the next two decades.

6.2 Database and Methodology

6.2.1 Database

The present study employs multiple thematic data layers derived from reliable and authoritative sources, including OpenStreetMap (OSM), Landsat satellite imagery, topographical sheets, and Google Earth. Both spatial and non-spatial datasets form the core foundation of geospatial research and analysis. Among these, OpenStreetMap (OSM) represents a vital source of spatial data, offering geographically referenced information contributed by a global network of volunteers. It encompasses a wide range of geospatial elements such as roads, buildings, water bodies, parks, and landmarks, which serve diverse applications in urban planning, navigation, environmental assessment, and disaster management. As a collaborative, crowd-sourced platform, OSM has revolutionized the way geographic data is collected, updated, and shared, providing an open-access framework for spatial research.

For remote sensing analysis, multi-temporal satellite imagery from Landsat-5 TM and Landsat-8 OLI was acquired from the United States Geological Survey (USGS) Earth Explorer Portal (https://earthexplorer.usgs.gov). These datasets offer high-resolution temporal snapshots that enable the detection of land cover changes, monitoring of urban expansion, and assessment of environmental transformations across the study area with remarkable precision. The Landsat data's consistent spatial and radiometric quality makes it ideal for long-term urban ecological studies.

Complementing the satellite data, topographical sheets provide critical spatial details regarding elevation, contour patterns, and terrain morphology, which are essential for understanding landscape dynamics and supporting geospatial modeling. Google Earth, on the other hand, integrates both spatial and non-spatial data to create an interactive 3D representation of the Earth's surface, offering satellite imagery, digital elevation data, and locational information. It enhances visualization, facilitates ground verification, and assists in delineating geographic features with greater contextual accuracy.

In summary, the integration of datasets from OpenStreetMap, Landsat imagery, topographical maps, and Google Earth has furnished this research with a comprehensive and multi-dimensional database. These sources collectively ensure the accuracy, depth, and analytical robustness necessary for assessing the spatial patterns and vulnerabilities of urban green spaces within the study area.

6.2.2 Methods

6.2.2.1 Development of Indicators Based on PSR (Pressure-State-Response) Criteria

In this chapter, a systematic and scientific methodology was adopted for the formulation of indicators under the Pressure-State-Response (PSR) framework. The process began with the identification of 15 preliminary indicators, derived from an extensive literature review and expert consultations (Gao et al., 2015; Bardhan et al., 2016; Lu et al., 2016; Chakraborti et al., 2018; Wu et al., 2020; Dinda et al., 2021; Ghosh et al., 2021; Dinda et al., 2022). These initial indicators were then subjected to a rigorous screening process, taking into account the geo-physical and environmental characteristics of the study area, as well as additional expert feedback. Based on this evaluation, a final set of 10 refined indicators was selected for detailed analysis. The expert-assigned weights for each indicator were averaged to establish their relative importance within the PSR model. In this framework, "Pressure (P)" represents the anthropogenic and environmental stressors exerted on urban green spaces, "State (S)" denotes the existing environmental condition and quality of these spaces, and "Response (R)" refers to the adaptive and management measures addressing these pressures and their impacts. Out of the 10 selected indicators, six correspond to Pressure, two to State, and two to Response. The details of these parameter-based indicators, along with their environmental significance and data sources, are summarized in Table 6.1.

6.2.2.2 Preparation of Indicators

The following indicators were developed and utilized within the Analytic Hierarchy Process (AHP) model framework to evaluate the vulnerability of urban green spaces (Fig. 6.1).

(a) **Land use/land cover change detection method**

To examine spatio-temporal transformations in land use and land cover (LULC) over a 30-year period, the values obtained for distinct LULC categories were analyzed to determine the rate of change within each class (Puyravand, 2003; Sahana et al., 2018). The rate of change (R) was computed using the formula where C_1 and C_2 represent the area under each LULC category at two time intervals, t_1 (2020) and t_2 (1990), respectively. This approach facilitates a quantitative understanding of land cover dynamics and helps identify regions experiencing significant transformations in green cover and built-up development.

(b) **NDVI calculation**

The Normalized Difference Vegetation Index (NDVI) was applied to assess the extent and condition of vegetation cover across the study area. As a widely recognized

Table 6.1 PSR framework-based indicators for urban green space vulnerability assessment

Parameters	Indicators	Significance measurements	Data source
Pressure	Proximity to settlement	Distance from high-density built-up area, in meter. Green space adjacent to built-up area is more likely to be converted into a built-up area	OpenStreetMap and Geographic Information System Database
	Proximity to roads	Distance from major roads, in meter. Green space adjacent to the major roads is more likely to be vulnerable	OpenStreetMap and Geographic Information System Database
	Proximity to railways	Distance from railways and stations, in meter. There is a strong possibility of future vulnerability of UGS in the areas near the railways lines and stations	OpenStreetMap and Geographic Information System Database
	Proximity to streams	Distance from streams, in meter. There is a strong possibility of future vulnerability of UGS in the areas near the streams	OpenStreetMap and Geographic Information System Database
	Elevation	High elevation having areas are likely less vulnerable than low elevation having areas in respect to UGS due to flood inundation the study area because of its situation on flood prone area	United State Geological Survey and DEM Data processed
	Slope	Steep slope areas are more likely to vulnerable to the UGS than gentle slope area especially near the Ramsar recognized East Kolkata Wetland	United State Geological Survey and DEM Data processed
State	LULC	Availability of land use land cove classes in respect to the total area of the city and their future dynamism probability	Prepared LULC map based on Landsat database derived from United State Geological Survey
	NDVI	Normalized difference vegetation index gives an overview about the nature and status of green spaces	Prepared NDVI map based on Landsat database derived from United State Geological Survey

(continued)

Table 6.1 (continued)

Parameters	Indicators	Significance measurements	Data source
Response	LST	Land surface temperature gives an overview regarding the current thermal status of the city and also future probability of dynamism	Prepared LST map based on Landsat database derived from United State Geological Survey
	VCI	Vegetation condition index represent the present status of the water availability of surface and also the condition of the vegetation	Prepared VCI map based on Landsat database derived from United State Geological Survey

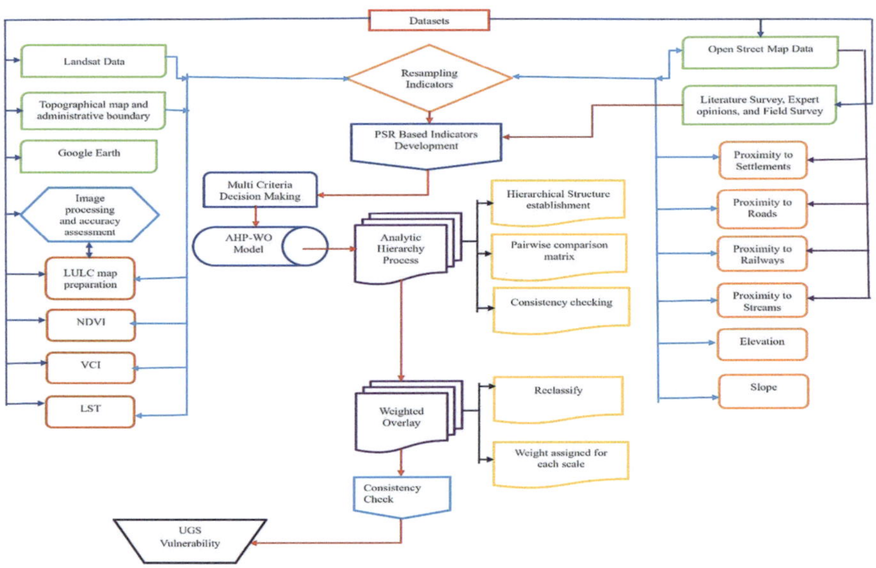

Fig. 6.1 Methodological flowchart for present UGS vulnerability assessment

vegetation index derived from satellite data, NDVI serves as an essential tool for monitoring vegetation health and spatial distribution (Townshend & Justice, 1995). It is especially effective for detecting and mapping temporal variations in vegetation vigor across large regions (Newnham, 2011). NDVI values range from -1 to $+1$, where higher values close to $+1$ indicate dense, healthy vegetation, while values approaching -1 represent barren or non-vegetated surfaces (Bardhan et al., 2016). The NDVI was calculated using the formula:

$$\text{NDVI} = (\text{Near Infrared (NIR)} - \text{Red}) / (\text{Near Infrared(NIR)} + \text{Red}).$$

(c) **Vegetation Condition Index (VCI)**

The Vegetation Condition Index (VCI) is a drought-monitoring indicator used to evaluate the current vegetation status relative to historical norms. It compares present NDVI values with long-term seasonal averages to assess vegetation stress and health (Liou & Mehabie, 2019). Expressed as a percentage, VCI was calculated using the equation:

$$\text{VCI} = ((\text{NDVIj}-\text{NDVImin})/(\text{NDVImax}-\text{NDVImin})) \times 100$$

where NDVIj denotes the current year's NDVI, NDVImax represents the multi-year maximum NDVI, and NDVImin corresponds to the multi-year minimum NDVI. This index effectively identifies deviations from normal vegetation conditions, providing insights into environmental stress and ecosystem resilience within urban green spaces.

6.2.2.3 Analytic Hierarchy Process (AHP) and Assigning Weight for Indicators

The Analytic Hierarchy Process (AHP), introduced by Thomas L. Saaty in the 1970s, is a widely recognized multi-criteria decision-making (MCDM) technique designed to solve complex problems through systematic evaluation and prioritization. It enables researchers to analyze multiple factors by structuring them into a hierarchical framework and determining their relative importance through pairwise comparisons. The AHP methodology involves three essential steps: (i) establishing a hierarchical structure of the problem, (ii) constructing a pairwise comparison matrix, and (iii) performing a consistency check to validate the assigned weights (Dyson, 2017; Dinda et al., 2022).

A nine-point fundamental scale (1–9) is utilized for pairwise comparisons, where each value reflects the relative importance of one indicator over another. This matrix facilitates the assessment of the relative priority and weightage of each indicator in the decision-making process. The mathematical representation of the pairwise comparison matrix is expressed as:

$$X = \begin{pmatrix} x_1 & x_2 \cdots & x_3 \\ x_8 & x_9 \cdots & x_{10} \\ x_{n1} & x_{n2} \cdots & x_{nn} \end{pmatrix}$$

$$P_{ij} = \frac{S_i}{S_j},$$

where P refers to importance of indicators, n is the total number of indicators. S_i denotes the weight of the indicators of alternative j. Consistency index was evaluated as:

$$CI = \frac{\lambda max - n}{n-1}$$

λ = maximum eigen value (from pairwise comparison matrix)

n = Total number of classes.

Matrix consistency was determined using consistency ratio (CR) as:

$$CR = \frac{CI}{RI'}$$

CR = consistency ration, CI = consistency index, and RI = ration index.

CR value equals or smaller to 0.1 or 10% is acceptable. CR value greater than 0.1 or 10% is subjected to review for assigning judgements.

Upon assigning appropriate weights to each indicator based on the AHP-derived rankings, these weighted parameters were integrated to develop the final Green Space Vulnerability (GSV) map, representing the spatial distribution of vulnerability across the study area (Fig. 6.1).

6.2.2.4 Weighted Overlay (WO) Model

The Weighted Overlay (WO) model is an essential analytical approach used in multi-criteria decision-making (MCDM) to address complex spatial problems by assigning rational and relative weights to multiple indicators. It is particularly effective in evaluating interrelationships and interdependencies among diverse parameters influencing a specific phenomenon (Liu et al., 2014). Within this model, each indicator is first reclassified on a raster layer, standardizing its values for analysis. All raster layers are then reclassified onto a common suitability scale, ensuring comparability across different parameters. These reclassified layers are subsequently overlaid, where each raster cell's suitability score is multiplied by the corresponding layer weight. The cumulative weighted values yield an overall suitability index, which is assigned to new cells in the output raster layer. The symbology of this output layer is then designed based on these composite values, providing a clear visual representation of spatial variations. In this study, each parameter was categorized into five equally spaced classes ranging from 0 to 5, representing different levels of vulnerability or suitability. These classes are expressed as $C = (C_1, C_2, C_3, C_4, C_5)$, where C denotes the evaluation categories corresponding to "Very High," "High," "Medium," "Low," and "Very Low" levels. This approach enables the integration of multiple weighted factors to produce a comprehensive Green Space Vulnerability Index (GSVI) map, reflecting the spatial distribution of vulnerability across the study area.

6.2.2.5 Vulnerable Zone Assessment of Urban Green Spaces

The identification of vulnerable zones both in the present and future is crucial for devising effective management strategies and implementing sound policy measures aimed at the conservation and sustainable management of Urban Green Spaces (UGS). In this study, the assessment of vulnerability was conducted by integrating indicators derived from the Pressure-State-Response (PSR) framework with Land Use and Land Cover (LULC) projections generated through the CA–Markov model, enabling a comprehensive evaluation of both current and anticipated future vulnerability patterns across the study area. To delineate the existing vulnerable zones, the researcher employed the Analytic Hierarchy Process (AHP) combined with the Weighted Overlay (WO) technique. This approach integrates multiple risk factors represented in raster format, each assigned a specific criteria weight determined using the AHP methodology. The integration of these spatial layers was carried out using the "Union" operator function in ArcGIS 10.5 (Nath et al., 2020), producing a composite spatial vulnerability map. For projecting future vulnerability zones of urban green spaces, a Multi-layer Perceptron (MLP) integrated CA-Markov chain model was applied, allowing for the prediction of spatial transitions and identification of potential areas at risk due to ongoing urban expansion and environmental pressures.

6.2.2.6 Image Processing and Land Use Land Cover Classification

The satellite images utilized in this study possessed a spatial resolution of 30 m for each selected year, following a consistent path (138) and row (44) configuration. Images captured during the spring season were exclusively chosen, as this period typically exhibits minimal atmospheric disturbances in the study area, ensuring higher data reliability and image clarity. The selected images underwent a series of preprocessing steps, including geo-referencing and rectification, to enhance spatial accuracy. They were projected using the UTM Zone 45N projection system based on the WGS 84 datum, ensuring uniformity in spatial alignment. Image registration was performed through the nearest neighbor resampling technique, incorporating 25 Ground Control Points (GCPs) to achieve optimal overlay precision (Ganaie et al., 2020).

To facilitate visual interpretation and training data selection, False Color Composites (FCC) were generated using appropriate band combinations. These composites aided in identifying and delineating training samples for each Land Use Land Cover (LULC) category. A total of 200 training sample points were collected per class, focusing on major land cover types such as built-up areas, water bodies, fallow lands, and green spaces. The training datasets were cross-validated using Google Earth imagery and topographical maps to ensure accuracy.

The classification process employed a supervised classification technique, specifically utilizing the Maximum Likelihood Classifier (MLC) algorithm, which is recognized for its statistical robustness in handling multi-spectral datasets. This methodology was applied to satellite images from two distinct time periods to assess temporal changes in land use and land cover. All image processing, classification, and map generation were conducted using ERDAS IMAGINE 2014 and ArcMap 10.5, ensuring a precise and systematic workflow for LULC analysis.

6.2.2.7 Multi-Layer Perceptron Model

The Multi-layer Perceptron (MLP) model, originally introduced by Rumelhart et al., (1986), is a widely adopted approach in remote sensing applications due to its exceptional capability to handle complex, nonlinear relationships and its strong generalization ability. In this study, the MLP model has been exclusively utilized for modeling and predicting Land Use and Land Cover (LULC) changes. Structurally, the MLP consists of three or more interconnected layers, an input layer, one or more hidden layers, and an output layer. The input layer functions as a channel for feeding data (such as spectral bands and auxiliary variables) into the network, without performing any processing. The hidden layers contain the computational nodes (neurons) responsible for internal processing, where the number of hidden layers and neurons is determined by the researcher. Finally, the output layer generates classification results corresponding to different land cover categories.

In this research, suitability maps for various LULC classes produced through the MLP were used to simulate future LULC conditions for the years 2030 and 2040. The modeling process comprised three main stages: (i) classification using a decision tree approach, (ii) modeling the transformation potential of each land cover class through MLP, and (iii) simulating future LULC dynamics using the Markov Chain model. LULC change maps were derived from time-series Landsat TM and OLI images, identifying the transition probabilities among different classes. These maps were integrated with multiple explanatory variables including elevation, slope, aspect, distance to degraded land, road, water bodies, settlements, and evidence likelihood to model transformation potential using the MLP framework. Unlike traditional parametric models, the MLP classifier offers robustness even with limited training data, effectively minimizing subjectivity in the analysis.

The Markov Chain transformation matrices were subsequently used to forecast the probability of future land cover transitions. However, while the Markov Chain method alone is limited in capturing spatial distribution patterns, its combination with MLP effectively models both the spatial and temporal dynamics of land cover change. This hybrid MLP–Markov Chain approach has proven particularly effective in modeling processes such as urban expansion and human-induced land degradation, producing highly reliable and spatially explicit results (Eastman, 2016). Model accuracy was validated by assessing the overlap between real and simulated maps, with an overlap exceeding 75% considered satisfactory. In cases where this threshold was not met, calibration procedures and refinement of environmental variables within the MLP–Markov Chain framework were undertaken to enhance the precision and reliability of the final simulation outcomes.

6.2.2.8 Simulation of Land Use Land Cover Using CA-Markov Chain

This section of the study emphasizes the analysis of key parameter-based factors influencing the future potential of urban expansion within the study area. Based on findings from an extensive field survey, six primary parameters were identified as critical determinants of future urban growth: distance from built-up areas, distance from roads, distance from railways, slope, distance from streams, and elevation levels, as outlined in Table 6.2. Each of these parameters was systematically processed to ensure compatibility for spatial analysis. Initially, the datasets were resampled and reprojected using ArcGIS 10.5, ensuring consistent spatial resolution and coordinate reference. The processed files were subsequently imported into the TerrSet 2020 mapping software for advanced modeling and simulation tasks.

The Euclidean Distance technique was employed to calculate the proximity of each Land Use Land Cover (LULC) influencing parameter, thereby generating distance-based raster layers. In this context, proximity plays a pivotal role areas located closer to built-up regions, roads, railways, and streams generally exhibit a higher likelihood of conversion into urban land, given their accessibility and developmental pressure. However, the elevation parameter behaves differently. Field observations indicated that zones located at higher elevations are more prone to urbanization compared to low-lying regions. This trend is primarily linked to the flood-prone characteristics of the study area, which is situated within the Gangetic deltaic plain. Consequently, urban growth patterns are influenced not only by infrastructural proximity but also by topographical and hydrological conditions, which collectively determine the spatial trajectory of future urban expansion modeled through the CA-Markov Chain framework.

6.2.2.9 Accuracy Assessment and Validation of Maps

The accuracy assessment and validation process are crucial methodological steps designed to evaluate the precision of classified images by comparing them with ground truth data, which includes verified geographical coordinates and corresponding real-world attributes. In this study, the Double Window Flexible Pace Search method was employed due to its superior ability to determine the optimal optical threshold for classification accuracy (Chen et al., 2006). This approach helped in refining the classification outcomes by reexamining and correcting potential errors in land use and land cover (LULC) extraction. Ground truth verification was conducted using Ground Control Points (GCPs) derived from both topographical maps and Google Earth reference data of the study area. Specifically, the 2020

Table 6.2 Details of indicators for future scenario development of urban green spaces

Indicators	Significance measurements	Data source	Data type	Methods
Distance from Built-up	Distance from high-density built-up area, in meter. Areas adjacent to built-up area is more likely to be converted into a built-up area	OpenStreetMap and Geographic Information System Database	Raster	Euclidean distance
Distance from Roads	Distance from major roads, in meter. Areas adjacent to the major roads is more likely to be converted into built-up	OpenStreetMap and Geographic Information System Database	Vector, converted to raster	Euclidean distance
Distance from Rails	Distance from railways and stations, in meter. There is a strong possibility of future conversion of areas near the railways lines and stations	OpenStreetMap and Geographic Information System Database	Vector, converted to raster	Euclidean distance
Distance from Streams	Distance from streams, in meter. There is a strong possibility of future conversion of areas into habitable near the streams	OpenStreetMap and Geographic Information System Database	Vector, converted to raster	Euclidean distance
Distance from Slope	Gentle slope areas are more likely to be converted into built-up than the steep slope areas	United State Geological Survey and DEM Data processed	Raster	Euclidean distance
Elevation	High elevation having areas are likely more prone to conversion into artificial land than low elevation land due to flood inundation nature of the study area because of its situation on flood prone area	United State Geological Survey and DEM Data processed	Raster	Slope tool in ArcGIS

LULC data was validated using these GCPs to ensure consistency and accuracy in spatial classification.

The Overall Accuracy (OA) was calculated using the following formula:

$$\text{OA} = \frac{(\text{Number of True Positives} + \text{Number of True Negatives})}{\text{Total Pixels in Ground Truth}} \times 100\%.$$

Additionally, the Kappa Coefficient (κ), a discrete multivariate statistical measure proposed by Cohen (1960) and applied in remote sensing accuracy assessments (Foody, 1992; Kar et al., 2018), was utilized to evaluate the level of agreement between the classified and reference data. The Kappa statistic accounts for the agreement occurring by chance, making it a more robust indicator of classification accuracy. It is calculated as:

$$\kappa = \frac{(n \sum x_{ii} - \sum x_{i+} x_{+i})}{(n^2 - \sum x_{i+} x_{+i})},$$

where r denotes the number of matrix rows, x_{ii} represents observations in row i and column i, x_{i+} and x_{+i} are the marginal totals for rows and columns respectively, and n is the total number of observations (pixels).

Two satellite images, corresponding to 2010 and 2020, were classified into four major LULC categories: Built-up, Water Body, Fallow Land, and Green Space. To ensure the validity and reliability of these classifications, 200 training sample points were meticulously selected and verified using Google Earth imagery and topographical datasets. This rigorous validation process ensured that the final LULC maps achieved a high level of accuracy and reliability for subsequent spatial and temporal analyses.

6.2.2.10 Future Scenario Development

The development of future scenarios, particularly in relation to Land Use and Land Cover (LULC) projections, plays a vital role in understanding and anticipating the evolving spatial and environmental dynamics of the study area. Within the framework of this research, future scenario modeling has been undertaken for the years 2030 and 2040 with a specific focus on the conservation of urban green spaces (Fig. 6.2). This forward-looking approach provides valuable insights into potential patterns of urban expansion, sprawl, and the decline of green cover, thereby contributing to more informed urban management and planning strategies. The predictive modeling not only enhances comprehension of ongoing land transformation processes but also supports sustainable resource management, biodiversity preservation, and disaster risk mitigation. By projecting how land use changes might unfold over time, this analysis offers policymakers, planners, and researchers a scientific basis for strategic decision-making, enabling them to implement proactive measures that ensure a balanced coexistence between urban development and environmental sustainability in the future.

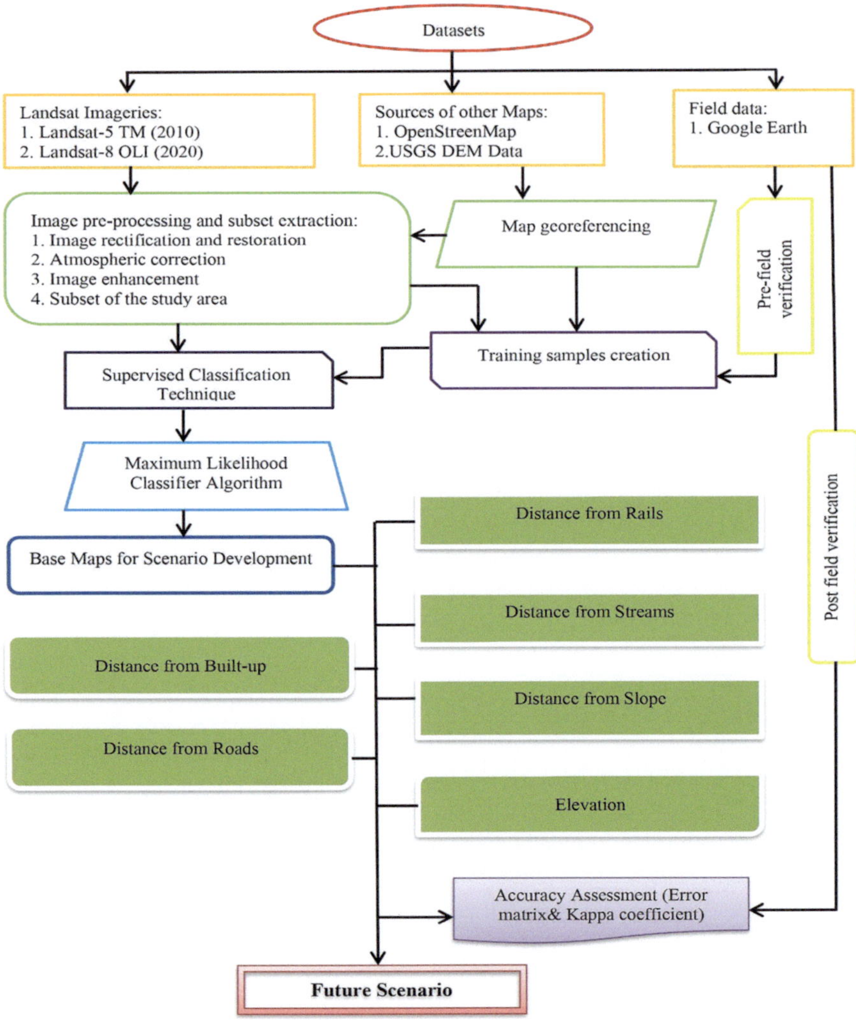

Fig. 6.2 Methodological flowchart for future scenario development

6.3 Results and Discussion

6.3.1 AHP-Based Weight Allocation for Indicators

The pairwise comparison matrix serves as a fundamental element in the multi-criteria decision-making (MCDM) framework, widely applied to evaluate and prioritize various indicators influencing urban green space vulnerability. It integrates both expert opinions and quantitative data to determine the relative importance of each indicator in relation to others, thereby establishing a structured hierarchy of significance. Analyzing this matrix helps to identify which factors exert the greatest influence on the vulnerability of urban green spaces within the study area.

The matrix encompasses ten indicators (Fig. 6.3), categorized under the Pressure-State-Response (PSR) framework: "Proximity to Settlement," "Land Use and Land Cover (LULC)," "Proximity to Roads," "Normalized Difference Vegetation Index (NDVI)," "Vegetation Condition Index (VCI)," "Proximity to Railways," "Proximity to Streams," "Land Surface Temperature (LST)," "Elevation," and "Slope." Each indicator is compared pairwise against all others to assess its relative significance using a numerical scale, where values range from 1 (equal importance) to 5 (much more important).

As observed in the pairwise comparison results (Table 6.3), certain indicators such as "Proximity to Settlement," "LULC," and "Proximity to Roads" consistently receive higher scores, highlighting their stronger influence on the pressure and vulnerability of urban green spaces. In contrast, indicators like NDVI, VCI, and LST receive relatively lower scores, suggesting that they contribute less directly to vulnerability in comparison. For instance, a value of 2 between "Proximity to Settlement" and "LULC" indicates that the former is considered twice as significant as the latter. Meanwhile, indicators such as "Elevation" and "Slope" consistently register equal importance, denoted by a value of 1, implying parity in their influence.

The normalized principal eigenvector column reflects the derived weights assigned to each indicator through the Analytic Hierarchy Process (AHP) computation. These weights quantify the comparative importance of each factor in evaluating urban green space vulnerability. The Consistency Ratio (CR), calculated at 6%, indicates a high level of reliability and coherence in expert judgments, as values below 10% are deemed acceptable (Saaty, 1980).

Overall, the pairwise comparison matrix acts as a structured decision-support tool, enabling policymakers and planners to identify the most influential parameters shaping the vulnerability of urban green spaces. Indicators with higher weights demand priority in conservation and management efforts, while those with lower weights may be addressed secondarily. Consequently, this AHP-based approach provides a transparent and scientifically grounded framework for urban green space policy formulation, ensuring that interventions align with environmental sustainability and urban resilience goals.

6.3.2 PSR-Based Indicators Weight

Table 6.4 presents the assigned weights of various indicators categorized under the Pressure-State-Response (PSR) framework, emphasizing their relative significance in assessing the vulnerability of urban green spaces. Within this framework, indicators are systematically grouped into pressure, state, and response components, each representing a different dimension of environmental influence and resilience. The assigned weights quantify the degree of importance attributed to each indicator in

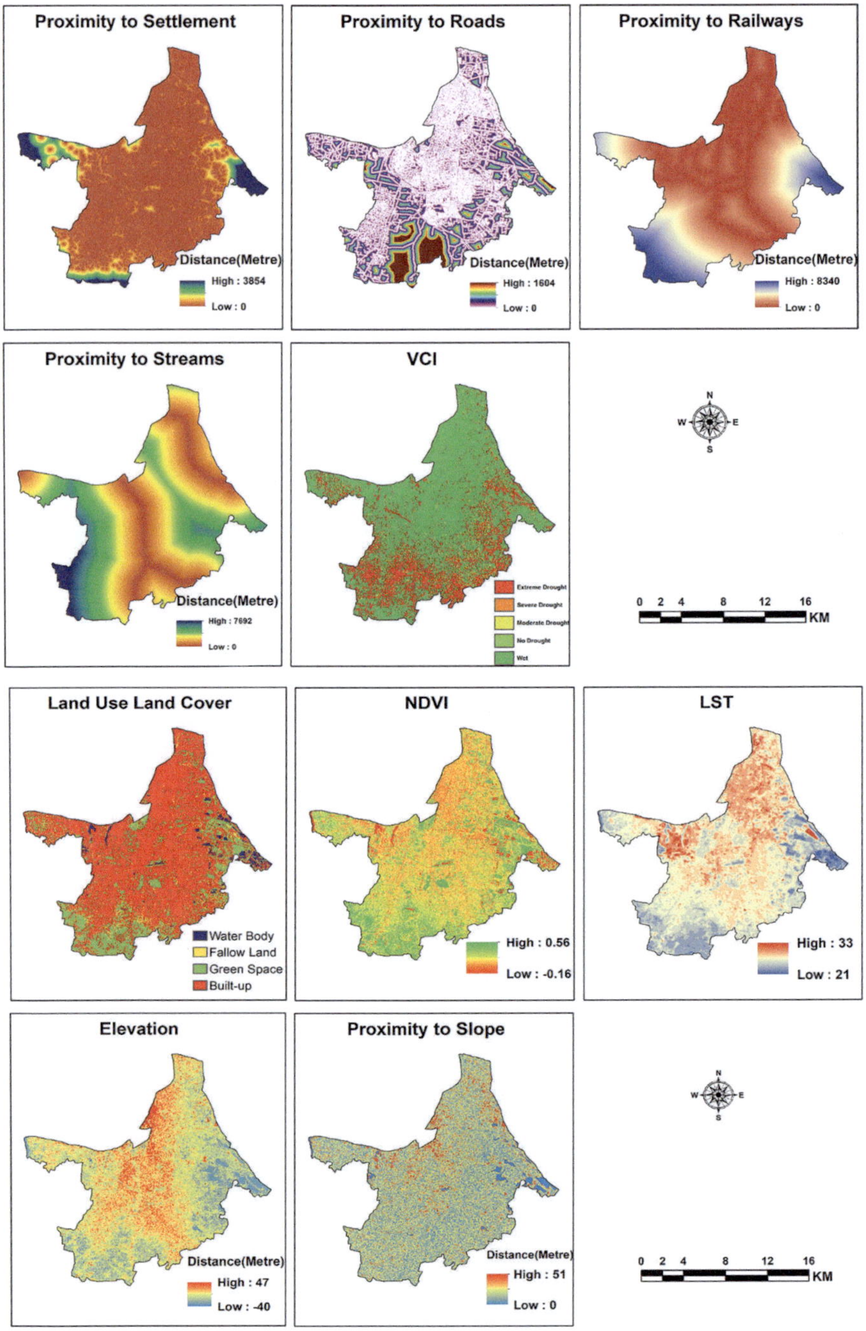

Fig. 6.3 PSR framework-based indicators for the assessment of urban green space vulnerability

Table 6.3 Pairwise comparison matrix for indicators

Matrix		Proximity to Settlement	LULC	Proximity to Roads	NDVI	VCI	Proximity to Railways	Proximity to Streams	LST	Elevation	Slope	Normalized principal Eigenvector (%)
		1	2	3	4	5	6	7	8	9	10	
Proximity to Settlement	1	**1**	1	1	3	3	3	3	5	1	1	17.0
LULC	2	1	**1**	1	3	2	1	1	5	1	1	13.0
Proximity to Roads	3	1	1	**1**	3	3	1	3	5	1	1	15.0
NDVI	4	0.33	0.33	0.33	**1**	1	0.33	0.33	1	1	1	5.0
VCI	5	0.33	0.5	0.33	1	**1**	0.33	0.5	1	1	1	5.0
Proximity to railways	6	0.33	1	1	3	3	**1**	3	3	1	1	13.0
Proximity to streams	7	0.33	1	0.33	3	2	0.33	**1**	3	1	1	9.0
LST	8	0.2	0.2	0.2	1	1	0.33	0.33	**1**	1	1	5.0
Elevation	9	1	1	1	1	1	1	1	1	**1**	1	9.0
Slope	10	1	1	1	1	1	1	1	1	1	**1**	9.0

Consistency Ratio CR = 6%

The bold values denote indicators that exhibit comparatively higher relative importance within the Analytical Hierarchy Process (AHP) framework. Specifically, bold formatting is used to highlight those indicators that have greater influence in the pairwise comparison matrix and consequently contribute more substantially to the normalized principal Eigenvector (final weights). These values represent criteria that exert a strategically dominant role in decision-making, reflecting stronger expert judgment preferences during pairwise evaluations

Table 6.4 PSR-based indicators weight

Criterion		Weights (%)
1	Proximity to settlement	17.0
2	LULC	13.0
3	Proximity to roads	15.0
4	NDVI	5.0
5	VCI	5.0
6	Proximity to railways	13.0
7	Proximity to streams	9.0
8	LST	5.0
9	Elevation	9.0
10	Slope	9.0
	Total	100.0

influencing urban green space conditions, and the cumulative total of all weights equals 100%, reflecting a complete distribution of importance across all parameters (Fig. 6.4).

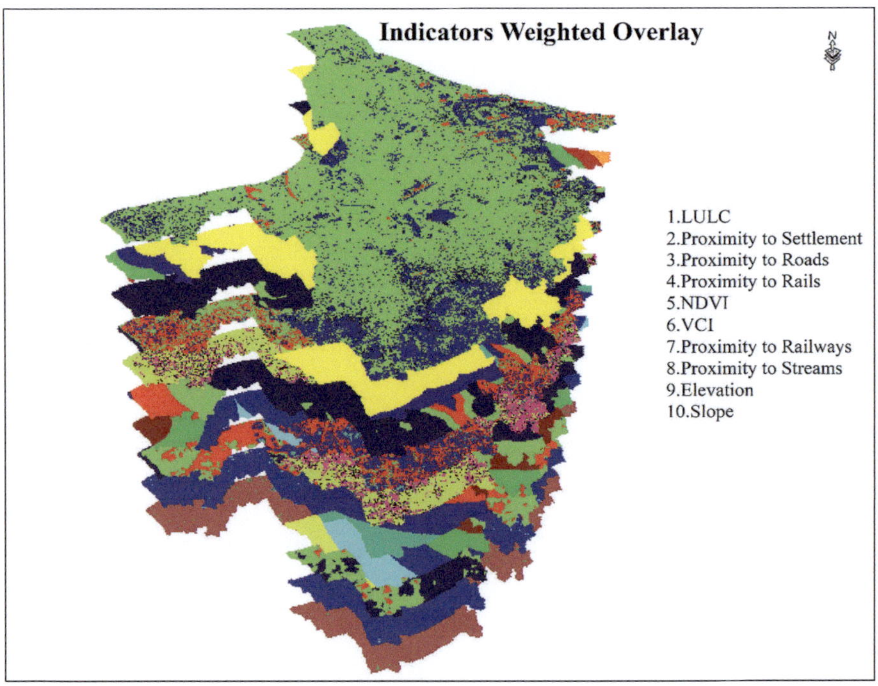

Fig. 6.4 PSR-based indicators weighted overlay

Among the evaluated indicators, "Proximity to Settlement," "Land Use and Land Cover (LULC)," and "Proximity to Roads" emerge as the most influential, carrying weights of 17.0, 13.0, and 15.0%, respectively (Table 6.4). These high-weight indicators represent the primary sources of pressure on urban green spaces. "Proximity to Settlement" captures the extent of anthropogenic impact, "LULC" reflects the degree of land conversion and urban expansion, while "Proximity to Roads" highlights potential fragmentation of green habitats and increased urbanization pressures. Following these, "NDVI" and "VCI" both assigned 5.0% serve as indicators of the state of vegetation health and density, reflecting ecological vitality but exerting relatively less pressure compared to the preceding parameters.

Other indicators, such as "Proximity to Railways" (13.0%), "Proximity to Streams" (9.0%), "Land Surface Temperature (LST)" (5.0%), "Elevation" (9.0%), and "Slope" (9.0%), are also crucial but contribute moderately to the overall vulnerability assessment. These parameters primarily represent secondary environmental and topographical influences that indirectly affect the sustainability of urban green spaces.

From a policy perspective, interpreting these weights is essential for prioritizing strategic interventions. Indicators with higher weights, particularly Proximity to Settlement, LULC, and Proximity to Roads, demand focused planning, conservation measures, and urban regulatory frameworks to mitigate their impacts. Conversely, while the lower-weight indicators play supportive roles in maintaining ecological balance, they may be addressed through broader environmental management policies. In summary, the PSR-based weighted analysis provides a structured and quantitative foundation for decision-makers to design targeted strategies for sustaining and protecting urban green spaces within the study area.

6.3.3 Urban Green Spaces Vulnerability

Table 6.5 presents an evaluation of the vulnerability levels of urban green spaces (UGS), classifying them into five distinct categories that reflect varying degrees of susceptibility to environmental and anthropogenic stressors. The percentage distribution highlights the proportion of UGS falling within each vulnerability class, thereby offering an essential framework for assessing the ecological sensitivity and resilience of these urban ecosystems. This categorization facilitates a comprehensive understanding of spatial variations in vulnerability and aids in identifying critical areas requiring immediate attention for conservation and management interventions (Fig. 6.5).

According to Table 6.5, the "Very Highly Vulnerable" category comprises 6.28% of the total UGS area, indicating regions that are under severe stress and face a significant risk of environmental degradation. These areas demand urgent and targeted management efforts to mitigate the adverse effects of urban pressures and restore ecological balance. The "Highly Vulnerable" category, accounting for the largest share at 81.84%, represents the majority of UGS within the study area. Such a high

Table 6.5 Urban green space vulnerability

Category wise vulnerability	Area (%)
Very highly vulnerable	6.28
Highly vulnerable	81.84
Moderately vulnerable	7.83
Low vulnerable	4.06
Very low vulnerable	0
Total	100

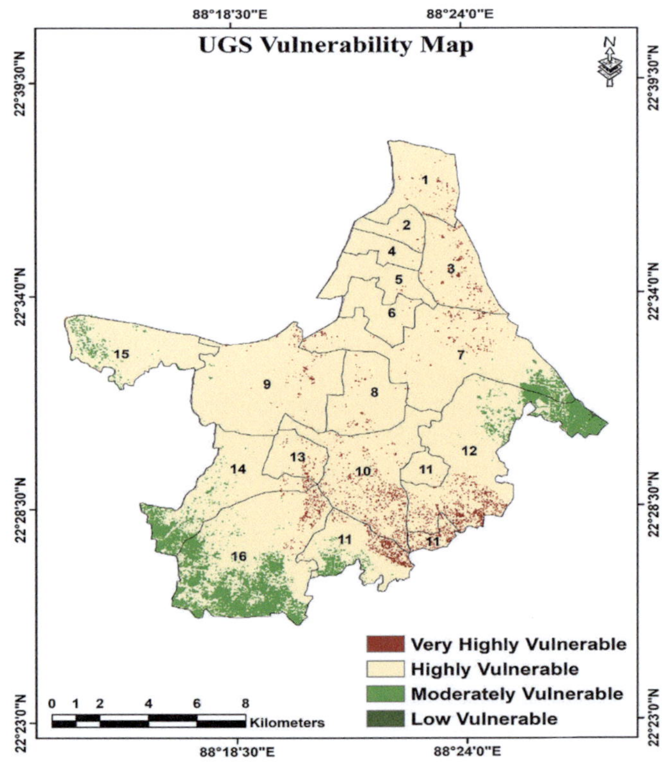

Fig. 6.5 Urban green space vulnerability map

percentage underscores the extensive exposure of green spaces to multiple urban stressors such as encroachment, pollution, and infrastructural expansion highlighting the necessity for immediate policy interventions to enhance their resilience.

The "Moderately Vulnerable" class includes 7.83% of UGS, reflecting zones that, while less threatened, still face moderate environmental stress. Timely implementation of preventive measures and sustainable management practices in these areas could prevent further deterioration. The "Low Vulnerable" category covers 4.06% of UGS, signifying areas with relatively stable ecological conditions, often located in less urbanized zones with natural buffering features. However, these areas should not be neglected, as rapid urban growth could increase their vulnerability over time. Notably, there are no UGS areas classified under "Very Low Vulnerable," indicating that all green spaces in the study area are affected to some degree by urban pressures, a reflection of the city's intense developmental dynamics and limited ecological stability.

From a policy and planning perspective, the findings emphasize the need for prioritized interventions in the "Very Highly" and "Highly Vulnerable" zones, including increasing vegetation cover, reducing pollution sources, and integrating green infrastructure into urban design. Meanwhile, "Moderately" and "Low Vulnerable" areas can serve as benchmarks for developing adaptive strategies and replicating resilient ecological models across more threatened zones. Overall, this vulnerability assessment offers a data-driven foundation for formulating sustainable urban green space management strategies that balance urban development with environmental preservation.

6.3.4 Future Scenario Development

Projecting future scenarios by incorporating variables such as distance from built-up areas, roads, railways, streams, slopes, and elevation levels (Fig. 6.6) enables policymakers and planners to anticipate zones that are at high risk of urban encroachment and environmental degradation. Such predictive modeling allows for proactive management of urban expansion, facilitating the implementation of targeted conservation measures to protect ecologically sensitive regions while simultaneously promoting the expansion and preservation of urban green spaces. Within the framework of Land Use and Land Cover (LULC) modeling, scenario prediction serves as a crucial tool for urban planners, conservationists, and decision-makers, empowering them to design urban growth patterns that maintain a balance between developmental needs and environmental sustainability.

Figures 6.7 and 6.8 illustrate the predicted transformations in LULC categories between 2020 and 2040, presenting a comprehensive overview of anticipated landscape changes over the next two decades. Correspondingly, Table 6.6 provides quantitative data on the absolute and percentage distribution of these categories for both years, along with their projected changes. The results reveal a distinct pattern of urban intensification accompanied by a sharp decline in natural and semi-natural land cover types.

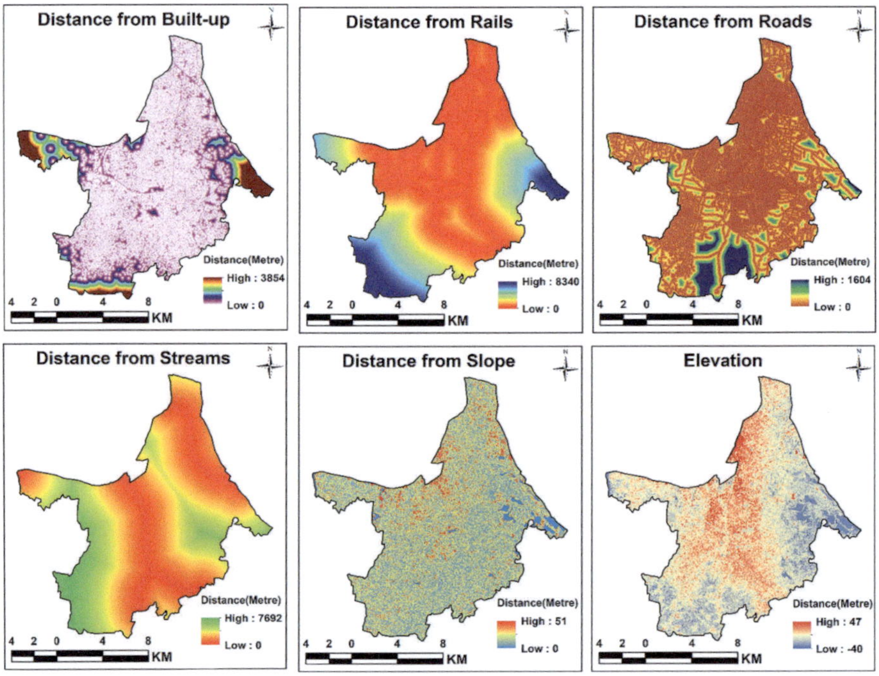

Fig. 6.6 Indicators for future scenario development of urban green spaces

According to Table 6.6, the built-up area is projected to expand significantly from 13,199.29 ha in 2020 to 19,328.35 ha in 2040, marking an increase of 6129.06 ha (46.43%). This substantial growth indicates rapid urbanization and infrastructural development, primarily at the expense of other land cover classes. Conversely, water bodies are expected to decline drastically from 1448.35 ha in 2020 to just 395 ha in 2040, representing a reduction of 1053.35 ha (− 72.73%) suggesting possible encroachment, pollution, or land reclamation activities. Similarly, fallow land is predicted to shrink from 675.48 to 511 ha, reflecting a 24.35% reduction, likely due to its conversion into built-up zones.

Most concerning is the projected decline in green spaces, which are expected to decrease from 5661.23 ha in 2020 to a mere 750 ha in 2040, representing an alarming 86.75% reduction. This sharp decline underscores the escalating threat posed by unplanned urban growth and highlights the urgent need for sustainable land management practices. Collectively, these projections demonstrate the pressing challenge of balancing urban expansion with environmental preservation.

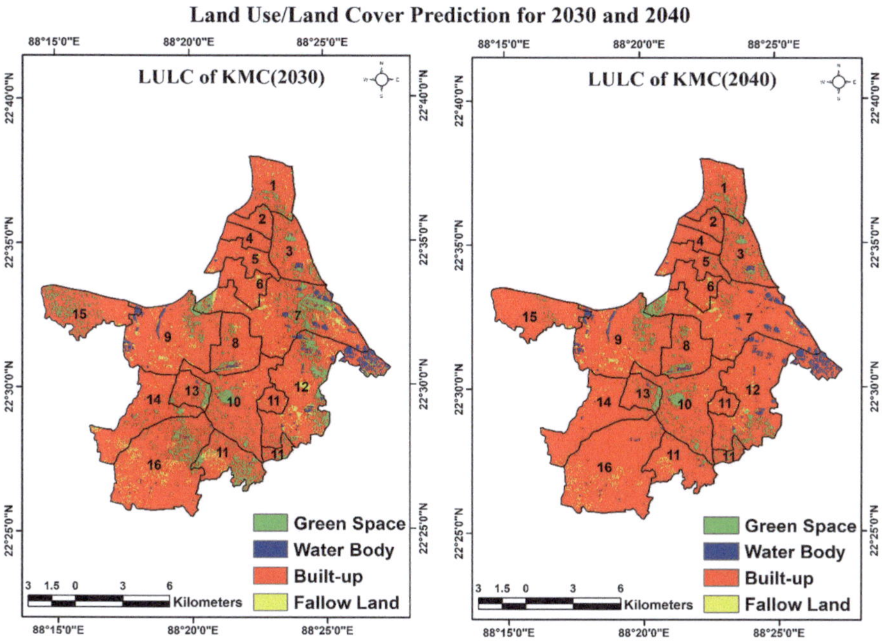

Fig. 6.7 Future scenario up to 2040

The observed trends underscore the importance of adopting integrated urban planning approaches that emphasize green infrastructure, ecological zoning, and policy interventions aimed at conserving remaining green spaces. Through informed planning and adaptive management, cities can mitigate the adverse impacts of urbanization and foster resilient, sustainable, and livable urban environments for the future.

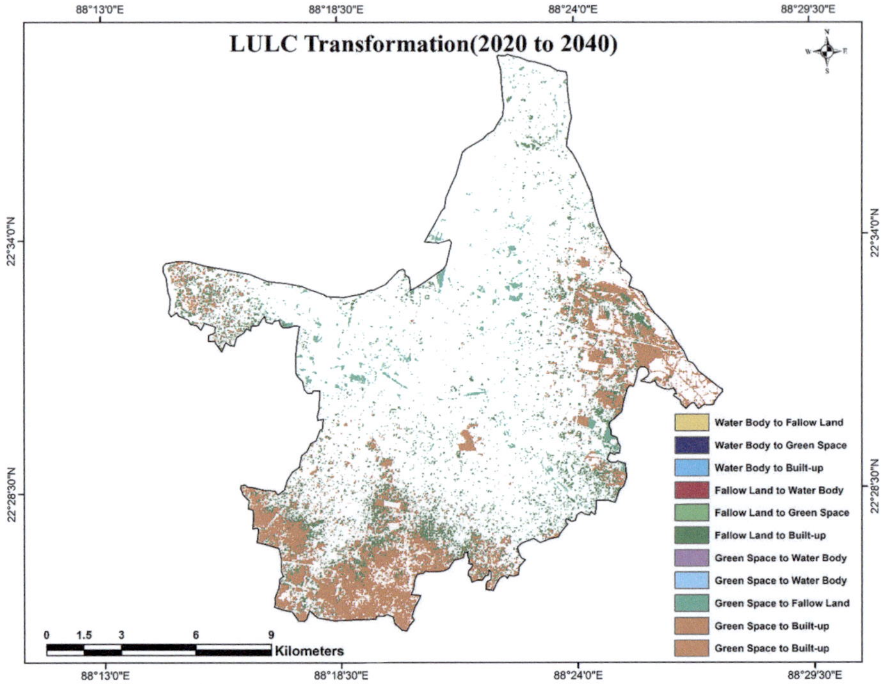

Fig. 6.8 Transformation prediction up to 2040

Table 6.6 Future scenario up to 2040

LULC Category	2020		2040		Changes (2020–2040)	
	Ha	%	Ha	%	ha	%
Built-up area	13,199.29	62.90	19,328.35	92.11	6129.06	46.43
Water body	1448.35	6.90	395	1.88	– 1053.35	– 72.73
Fallow land	675.48	3.22	511	2.44	– 164.48	– 24.35
Green space	5661.23	26.98	750	3.57	– 4911.23	– 86.75
Total	20,984.35	100.00	20,984.35	100.00		

Chapter 7
Conclusion and Policy Recommendations

7.1 Comprehensive Summary of the Key Findings Derived from the Study

Urban green spaces act as essential "lungs of the city," offering not only environmental and material benefits but also serving as vital reservoirs of biodiversity. In urban regions like the Kolkata Municipal Corporation (KMC), these spaces deliver a wide array of ecological, social, physical, and economic advantages. They sustain a diverse range of flora and fauna within expanding urban landscapes, absorb carbon dioxide, mitigate the urban heat island effect, and enhance resilience against climate change. Moreover, they promote mental and physical well-being by reducing stress and encouraging active lifestyles. Thus, urban green spaces are fundamental to ensuring the sustainability, livability, and resilience of the city.

However, the continuous expansion of urban infrastructure and rapid population growth in Kolkata have caused a sharp decline in green areas, leading to environmental degradation and loss of biodiversity. The displacement of species and disturbance of ecological harmony due to unplanned development emphasize the urgent need for strategic measures aimed at conserving, improving, and expanding existing green spaces. Safeguarding biodiversity-rich zones within the city is vital for maintaining ecological balance and supporting native species in this densely populated metropolis.

The significant findings from the study are outlined below:

1. Over the three-decade period (1990–2020), KMC experienced substantial land use and land cover (LULC) transformation driven by urbanization and demographic expansion. The built-up area expanded by 45.13%, increasing from 9094.90 ha in 1990 to 13,199.29 ha in 2020, primarily in the southern, southeastern, and southwestern sectors. In contrast, green spaces declined by 2338.02 ha (29.23%), highlighting the strain on ecological balance. Population growth, including both permanent and floating populations, has been the

© The Author(s), under exclusive license to Springer Nature Switzerland AG 2026
M. B. Ali and S. Jamal, *Urban Green Spaces and Environmental Sustainability*,
https://doi.org/10.1007/978-3-032-15322-7_7

major driver of these shifts. A strong positive correlation (92%) between population increase and built-up expansion underscores the need for integrated urban planning that harmonizes development with green space conservation.

2. The Land Surface Temperature (LST) across KMC varied between 20 and 35 °C, indicating borough-level and temporal differences in thermal conditions. Some boroughs exhibited rising LST trends, while others showed stability or decline. Borough 7 consistently recorded the highest temperatures. These fluctuations reflect influences such as the urban heat island effect, LULC alterations, climate variability, and local mitigation initiatives. Using LST, NDVI, and NDBI indices, this study analyzed the spatio-temporal transformation of urban green spaces from 1990–2020. The findings illustrate how changes in vegetation and built-up intensity have significantly contributed to rising urban temperatures and altered microclimate patterns, supported by detailed ward-wise analysis.
3. Chapter four established a direct connection between green spaces, air quality, and noise pollution across KMC. Areas with limited greenery, such as Ballygunge and Bidhan Nagar, recorded PM2.5 and PM10 concentrations above permissible limits, while greener boroughs like Thakurpukur and Jadavpur exhibited comparatively cleaner air. Similarly, highly urbanized localities such as Rabindra Bharati University and Bag Bazar reported elevated pollution levels. In contrast, Victoria Memorial and the Maidan (SSKM Hospital), with abundant green cover, recorded significantly lower levels of air and noise pollution. Conversely, Gole Park and Bag Bazar, dominated by concrete surfaces, experienced high noise intensity. These findings confirm the crucial role of vegetation in mitigating pollution and enhancing urban environmental quality.
4. Based on primary survey data collected through the interview-cum-schedule method across 16 boroughs (covering 206 km^2), Chapter five analyzed the impact of green space on psycho-social and physical health. The study tested two hypotheses:

   - Hypothesis 1 revealed a strong positive correlation ($r = 0.675, p = 0.005^{**}$) between green space availability and psycho-social well-being, indicating that residents with greater access to green spaces exhibit higher mental wellness.
   - Hypothesis 2 found a moderate positive correlation ($r = 0.4$, $p = 0.02^{**}$) between green space and physical health, implying that an increase in green cover improves physical well-being. Negative correlations between distance to green spaces and both the frequency (– 0.219) and duration (– 0.351) of physical activities demonstrate that proximity to greenery encourages active lifestyles, reducing health risks and improving quality of life.

5. Chapter six synthesized insights from the application of the Analytic Hierarchy Process (AHP), the Pressure-State-Response (PSR) framework, vulnerability mapping, and future scenario modeling. AHP facilitated the weighting of indicators, identifying "Proximity to Settlement," "LULC," and "Proximity to Roads" as the dominant vulnerability drivers, whereas "NDVI," "VCI," and "LST" held lower significance. The PSR model classified indicators into pressure, state, and response categories, clarifying their contextual roles in influencing vulnerability.

The vulnerability assessment delineated high and very high vulnerable zones, warranting immediate policy attention, and moderate to low vulnerable zones, which offer opportunities for resilience strengthening. Future projections reveal escalating built-up expansion alongside dramatic declines in green spaces and water bodies. These results stress the necessity for sustainable land management, integrated planning, and policy-driven conservation to ensure the coexistence of urban development and ecological preservation, fostering a resilient and sustainable urban ecosystem for the Kolkata Metropolitan region.

7.2 Plans, Schemes, and Policies Implemented in Recent Years for the Conservation and Management of Urban Green Spaces in the Study Area

The Kolkata Municipal Corporation (KMC) utilizes data provided by CCCKolkata (Centre for Contemporary Communication) to plan and conserve the city's parks effectively. CCCKolkata is dedicated to promoting the urban biodiversity, ecosystem, and greening of Kolkata. Since 2012, they have successfully planted 10,000 trees in Kolkata, with a 75% survival rate. Today, these trees stand 25–30 feet tall and continue to grow. CCCKolkata maintains these trees for two years from the date of planting. The tree plantation initiative spans various locations, including Kolkata Maidan, Fort William, several city parks, community homes, schools, colleges, slums, Rabindra Sarovar National Park, and Subhash Sarovar. Collaborating partners include KMC, KMDA, PWD, corporate houses, educational institutions, eastern command at Fort William, and other NGOs.

The neighborhoods across the 16 boroughs and 144 wards of Kolkata boast numerous big and small urban neighborhood parks, totaling more than 750. These parks play a pivotal role in greening the city and tree planting. They serve as urban biodiversity hotspots, contributing to climate change mitigation and adaptation efforts. Additionally, they aid in reducing environmental pollution, particularly vehicular pollution, support groundwater recharge, and provide a source of oxygen for breathing. Moreover, these parks serve as common spaces where the diverse population of Kolkatans can come together, fostering a unified common identity and sense of community.

According to a World Bank Report, Kolkata is identified as one of the prominent cities vulnerable to climate change, facing the looming threats of destructive cyclones and sea storm surges over a span of a hundred years. CCCKolkata has been dedicated to working toward environmental sustainability and inclusive development in the city of Kolkata since 2010. Its efforts encompass research, publications, partnerships, advocacy, mobilization, and development communications. Starting in 2010, it initiated work to conserve urban greens and city parks, which has seen significant progress in the current financial year through partnerships with the Kolkata Municipal Corporation (KMC) and the Regional Remote Sensing Centre of ISRO (RRSC).

They continue to raise awareness by engaging with schools, colleges, and communities. To support their endeavors, they are actively seeking strategic collaborations with corporations to secure funding. The custodian of these parks and urban greens, KMC, unfortunately, lacks the budget and resources for their conservation and development. With nearly 750 urban parks under their care, KMC operates on a modest budget of approximately Rs. 38 crores and employs a staff of 200 individuals. To address this challenge, CCCKolkata is exploring corporate and institutional support and forging partnerships to ensure successful conservation efforts.

The recent World Bank Report emphasizes the importance of public–private partnerships for any urban revival or rejuvenation project. In response, CCCKolkata is diligently working to activate such partnerships and has been in discussions with various corporate houses to sponsor parks as part of mission.

To establish a connection with the community, CCCKolkata has devised a strategy to collaborate with institutions and local NGOs. By leveraging the grassroots-level connections and accessibility of these organizations, CCCKolkata can effectively and swiftly communicate with a broader and more dispersed community. Since January 2015, CCCKolkata has partnered with an NGO called EKTARA to raise environmental awareness among the community it serves.

CCCKolkata initiated a program focused on environment and hygiene in EKTARA, which is a school catering to girl children in Topsia, KMC ward number 59. This area is known to be one of the most polluted parts of the city. EKTARA primarily recruits its staff and teachers from the local community. CCCKolkata joins forces with EKTARA to promote environmental sustainability and inclusive development. The program addresses critical issues such as health, hygiene, solid waste management, urban greens, the significance of trees, and fostering personal activism for Kolkata's environmental well-being. The primary objective of this initiative is to provide education and action plans related to hygiene and the environment.

The preservation of Kolkata's parks and green spaces plays a crucial role in transforming the city into a livable, workable, and recreational hub. This project holds significant importance in making Kolkata a green city and securing a sustainable future for its residents amidst the challenges posed by climate change. Urgent action is required, as CCCKolkata's research indicates that rapid urbanization could lead to the loss of many of these parks, replaced by built-up areas or private spaces.

With technical assistance from RRSC-ISRO, GoI, CCCKolkata has mapped the GPS positions of the parks, which were collected through their survey, onto a satellite image of KMC. This information is now accessible through ISRO's Bhuvan platform to all residents of the city and beyond. Capitalizing on these findings, CCCKolkata is engaging in discussions with select corporations, particularly those in the builder community, to encourage their adoption of specific parks. This approach aims to find a viable solution for conserving the extensive network of approximately 750 parks. Additionally, CCCKolkata is striving to gather funds to establish a partnership with KMC for the conservation of all the parks in Kolkata.

The maintenance and ownership of these parks belong to the Kolkata Municipal Corporations (KMC), which serves as the city's Local Self-Governing (LSG) body.

Noteworthy steps and initiatives are focused on conserving and improving the green spaces throughout the city.

1. During April and May 2022, a significant effort has been made to plant 1000 trees in Green Verge 11, Action Area 2, New Town, Kolkata.
2. On December 22, 2021, the inauguration and public dedication ceremony for Verge 18 in New Town were held, benefiting the local neighborhood.
3. On August 18, 2021, CCC commenced its Kolkata City Tree Plantation 2021, resulting in the planting of 1100 trees in Rabindra Sarovar, Subhas Sarovar, and 2100 trees in New Town (verge 18, DC Block, New Town, opposite Tank Number 7/New Town School) during August to October 2021.
4. On September 1, 2020, the Centre for Contemporary Communication, in collaboration with Srijan Realty and KMDA, successfully planted 1000 trees at various locations in Kolkata, including Rabindra Sarovar, Subash Sarovar, and New Town. The plantation was completed in September 2020.
5. In October 2019, CCCKolkata partnered with Bengal Area, Eastern Command, and Srijan Realty to plant over 500 trees in Kolkata Maidan. Additionally, CCCKolkata planted 100 trees in Tiljala Basti, in association with Tiljala Shed.
6. Between September 26th and 29th, 2019, CCCKolkata, together with Eastern Command and Srijan Realty, planted 402 trees in Fort William.
7. On July 26, 2019, CCCKolkata joined forces with KMDA to plant 10 trees in Rabindra Sarovar campus. The tree plantation was inaugurated by the honorable minister, Shri Firhad Hakim, and involved various species like Karanja, Hijal, Jam, Sirish, Awastha, Sita Asok, Neem, Amalthas, Bakul, and Jarul. The plantation took place in a vacant space opposite the main gate of Nazrul Mancha.
8. In 2014, CCCKolkata planted 2000 trees in Kolkata Maidan, popularly known as the Lungs of Kolkata, established by the British almost 300 years ago. Prior to this project, there had been no scientific tree planting attempts at the Maidan. Siddha Realty, a prominent builder in the city, supported CCCKolkata's initiative by granting Rs. 100,000, providing 1000 tree guards (each costing Rs. 500), and donating 2000 tree saplings (each with a minimum height of 5 to 6 feet, costing at least Rs. 200). This effort involved the participation of twenty-five schools and their children, with nearly 250 children involved in planting saplings and installing tree guards. KMC collaborated with CCCKolkata and provided 1000 tree guards for the drive, as these public commons space required protection for each tree.

Addressing the environmental impacts of urban green spaces in Kolkata Municipal Corporation is of utmost importance. But the rapid urbanization and expansion of infrastructure in this region have resulted in a concerning decline in these green spaces, leading to environmental degradation and related challenges. It requires the development and implementation of comprehensive strategies to preserve and enhance existing green areas, create new ones, and integrate them into urban planning and development. Efforts should be made to identify and protect vulnerable hotspots within the city, ensuring the proper management plans for maintaining the ecological balance. It needs a holistic approach for their planning, conservation and management of the urban green spaces. Following are some crucial suggestions for the conservation and management of urban green spaces.

- It is essential to raise awareness among the citizens about the importance of urban green spaces and their environmental benefits. Community involvement should be encouraged through initiatives like community gardens, tree planting drives, and citizen science projects.
- Municipal authorities, urban planners, environmental experts, community organizations, and residents should work together to formulate and implement policies that prioritize the preservation and expansion of green spaces. Financial incentives, grants, and partnerships with private entities can also play a significant role in supporting these initiatives.
- As green space transformation in near future was observed by using CA-Markov Chain model, which necessitates the need for planning and management for conservation and expansion of green space by the governing authorities with collaboration of urban planners, ecologists, environmentalists and people participation.
- Green belt around the building, rooftop and balcony gardening is a very trendy concept at present time, in especially mega cities of developed countries. This could be a major initiative in present study area to make city greener.
- Green wall concepts and vertical green are another major steps to make the city green.
- VIP road is one of the widest and longest roads passing through the East Kolkata Wetland, therefore there is great scope to make roadside greening, which can be an important initiative by the urban planners in this study area.
- Along Hoogly River, other sub tributaries and canals have the great potential to develop green belt that could be the major step for greening the city.
- And mapping of vulnerable urban green space areas can help to adopt priority, implementation, and management policies.
- Green Infrastructure (GI) is donated as the natural or artificial vegetation and water area inside or around urban area, including forests, grasslands, lakes, rivers and parks, which could be a major step for urban greening. Therefore, there should be more focus by urban planners, ecologists and government to make more green infrastructure in the city.
- The KMC depicts, 80% of its green space come under vulnerable conditions, which require conservation and management on priority with strategic planning and management.
- Results indicate declining trend of UGS from 1990 to 2020 and likely further decline in the future. The loss of UGS should be managed properly by urban planners and governments for achieving sustainable urban growth.
- Contemporary environmental problems such as air pollution and urban heat island, which cause various health issues, to reduce such issues we need to increase the requirements of green space in the city by creating new green infrastructures.
- Urban expansion cannot stop but can be managed by following and implementing the "Urban and Regional Development Plans Formulation and Implementation

Guidelines" (MoUD, 2014) to protect the existing green space and consideration of plan to increase the areas of UGS to fulfill the minimum requirement.

7.3 Integrated Synthesis of the Role of Urban Green Spaces (UGS) in Promoting Environmental Sustainability

Urban Green Spaces (UGS) play a pivotal role in ensuring environmental sustainability, fostering ecological balance, and enhancing the overall health and well-being of urban populations. In rapidly urbanizing cities like those within the Kolkata Municipal Corporation (KMC), green spaces serve as essential buffers that mitigate the negative impacts of urban growth, pollution, and climate change. They are not merely aesthetic or recreational amenities but integral components of urban ecosystems that sustain biodiversity, regulate microclimates, and provide vital ecosystem services indispensable for human survival.

From an environmental sustainability perspective, UGS contribute significantly to improving air quality, reducing carbon footprints, and moderating urban heat. Trees and vegetation within parks, gardens, and roadside plantations act as natural air filters, trapping dust particles and absorbing harmful gases such as carbon dioxide, sulfur dioxide, and nitrogen oxides. This process helps to lower greenhouse gas concentrations and contributes to climate regulation at both local and regional scales. Green spaces also play an important role in temperature regulation, countering the urban heat island (UHI) effect by providing shade and facilitating evapotranspiration, which lowers ambient temperatures. In Kolkata, where dense built-up zones dominate, green corridors and urban parks act as micro-climatic regulators, offering much-needed thermal comfort during peak summer months. Additionally, vegetation cover enhances carbon sequestration, a key factor in achieving sustainable urban development goals (SDGs) related to climate action and sustainable cities.

From an ecological standpoint, UGS function as crucial habitats for urban biodiversity, supporting a variety of flora and fauna that contribute to the ecological resilience of cities. They maintain biological diversity by offering shelter, breeding grounds, and feeding sites for birds, insects, and small mammals, thereby sustaining ecological networks even within highly urbanized environments. In the context of the KMC area, larger green zones such as Rabindra Sarovar, Subhash Sarovar, and the Maidan serve as ecological hotspots that preserve native species and prevent local extinctions. Urban vegetation also improves soil quality, reduces surface runoff, and promotes groundwater recharge, thereby strengthening hydrological cycles. Moreover, integrating UGS within urban planning enhances ecosystem connectivity, forming "green corridors" that enable species migration and ecological continuity across fragmented landscapes.

The contribution of UGS extends beyond environmental and ecological dimensions to public health and well-being. Numerous studies have established a strong relationship between access to green spaces and improved physical, mental, and social health. Proximity to parks and gardens encourages physical activities such as walking, jogging, and cycling, reducing risks of lifestyle-related diseases such as obesity, cardiovascular disorders, and diabetes. Equally important are the psychological benefits, exposure to nature has been shown to lower stress levels, anxiety, and depression, while promoting emotional stability and cognitive restoration. For densely populated cities like Kolkata, where residents are often exposed to pollution and noise, urban greenery provides a therapeutic refuge, promoting community well-being and social cohesion.

Furthermore, UGS contribute to social sustainability by fostering interaction, inclusivity, and community participation. Parks and open spaces serve as communal gathering points, bridging social divides and encouraging a sense of belonging among diverse urban populations. They play a transformative role in creating inclusive urban environments where people from different socio-economic backgrounds share common spaces, thereby strengthening urban social fabric and civic engagement.

In an integrated perspective, UGS function as multi-dimensional assets that simultaneously address environmental, ecological, and social objectives. To achieve true urban sustainability, cities must adopt policies that protect, expand, and equitably distribute green spaces. Urban planning should embed green infrastructure within the broader framework of sustainable development aligning with global initiatives such as the Sustainable Development Goals (SDG 11 and SDG 13) to ensure resilient, livable, and climate-adaptive urban systems. In the case of Kolkata, strengthening the management of existing green spaces, restoring degraded ones, and ensuring equitable access across all boroughs are vital steps toward enhancing urban resilience and human well-being.

In conclusion, Urban Green Spaces are indispensable to the vision of sustainable, healthy, and resilient cities. Their ecological, climatic, and socio-health benefits underscore the necessity of integrating them as central pillars of urban governance and planning. The effective preservation, expansion, and scientific management of UGS will not only protect biodiversity and ecological functions but will also nurture healthier, more equitable, and sustainable urban communities for future generations.

7.4 Critical Assessment of Existing National and Local Policies Related to Urban Green Infrastructure, Land Use and Environmental Governance

Urban green infrastructure (UGI) is increasingly recognized as a critical component of sustainable urban development, integrating natural systems into the built environment to enhance ecological functions, social well-being, and climate resilience. In India, including the Kolkata Municipal Corporation (KMC) area, several national and local policies have been formulated to promote green infrastructure, sustainable land use, and effective environmental governance. However, despite progressive frameworks and institutional efforts, implementation challenges, policy fragmentation, and inadequate integration across administrative levels have limited their effectiveness. This section critically examines these policies, their strengths, and their gaps concerning urban environmental management and green space conservation.

At the national level, the National Urban Policy Framework (NUPF) and National Mission for a Green India (GIM) emphasize the restoration and enhancement of urban green cover. GIM, launched under the National Action Plan on Climate Change (NAPCC), aims to increase forest and tree cover and improve ecosystem services. While these initiatives recognize the ecological and social value of green spaces, they often prioritize forested and rural ecosystems over urban landscapes. The absence of specific urban green space guidelines limits their applicability in dense metropolitan areas like Kolkata. Similarly, the Atal Mission for Rejuvenation and Urban Transformation (AMRUT) and Smart Cities Mission (SCM) include components related to parks, urban forestry, and green belts. Under AMRUT, cities are encouraged to develop green spaces and parks to enhance urban livability. Kolkata has benefited from such schemes, yet projects often remain piecemeal, focusing on beautification rather than ecological functionality.

The Urban and Regional Development Plans Formulation and Implementation (URDPFI) Guidelines (2014) provide a robust framework for integrating green infrastructure in urban land use planning. They recommend allocating a minimum of 15–20% of city land for open and recreational spaces. However, enforcement remains weak, and rapid urban expansion often leads to the conversion of designated green areas into built-up zones. The Environmental Impact Assessment (EIA) Notification 2006 also mandates consideration of green buffers and landscaping in urban projects, but compliance and monitoring remain inconsistent, especially in informal settlements and peripheral urban zones.

At the local level, the Kolkata Municipal Corporation (KMC) has implemented several initiatives, such as the Kolkata Environmental Improvement Project (KEIP) and collaborations with organizations like CCCKolkata and KMDA, aimed at enhancing urban greenery and managing open spaces. However, KMC faces institutional and financial constraints, with limited budgets and manpower to maintain nearly 750 public parks under its jurisdiction. While KMC's by-laws acknowledge the importance of parks, wetlands, and open areas, their protection is often compromised by urban encroachments, infrastructure projects, and inadequate enforcement of land use regulations.

A critical issue lies in the fragmented governance structure. Multiple agencies such as KMC, KMDA, the State Forest Department, and the Urban Development Department share overlapping responsibilities for land management, leading to bureaucratic inefficiency and weak coordination. The lack of a unified policy framework for UGI management hampers data sharing, strategic planning, and cross-sectoral integration. Additionally, local-level participation in environmental governance remains minimal, with limited involvement of communities, resident welfare associations, and civil society in decision-making processes related to green space conservation.

Moreover, while policies like the National Biodiversity Action Plan (NBAP, 2014) and State Action Plans on Climate Change (SAPCC) emphasize ecosystem restoration and biodiversity enhancement, their translation into actionable urban strategies remains insufficient. In Kolkata, wetland degradation, tree loss, and land use conflicts persist, highlighting the gap between policy intent and implementation.

In conclusion, India's national and local policies provide a strong conceptual foundation for promoting urban green infrastructure and sustainable land use. However, the success of these policies depends on integrated governance, institutional coordination, and effective enforcement mechanisms. For cities like Kolkata, a dedicated Urban Green Infrastructure Policy integrated with land use planning, public health, and climate resilience frameworks is essential. Strengthening institutional capacity, encouraging public–private partnerships, and involving local communities in urban greening initiatives will ensure that policies translate into tangible, sustainable outcomes. Only through such coordinated and inclusive governance can urban green spaces become central to India's sustainable urban future.

7.5 Strategic Policy Recommendations Aimed at Fostering Sustainable Urban Planning

The growing pace of urbanization in India, particularly in metropolitan cities such as Kolkata, has created immense pressure on natural resources, including urban green spaces (UGS). These spaces, which serve as vital ecological buffers and public amenities, are increasingly under threat due to unregulated land conversion, infrastructure expansion, and inadequate policy enforcement. To achieve sustainable urban development, it is imperative to adopt a holistic and integrated policy approach that ensures sustainable urban planning, equitable access to green spaces, and long-term conservation of urban ecological assets. The following strategic policy recommendations outline a framework for enhancing the ecological and social sustainability of urban green spaces.

1. Integration of Green Infrastructure into Urban Planning Frameworks

A central policy priority should be the integration of urban green infrastructure (UGI) into all levels of city planning and development. Urban master plans and development plans must incorporate green corridors, ecological buffers, and open spaces as essential infrastructural components, rather than as supplementary amenities. Tools such as Geographic Information Systems (GIS) and Remote Sensing (RS) should be systematically used for mapping, monitoring, and managing green assets. The inclusion of green infrastructure in land use planning should follow the Urban and Regional Development Plans Formulation and Implementation (URDPFI) Guidelines, 2014, which recommend dedicating at least 15–20% of urban land to open and green spaces.

2. Strengthening Institutional Coordination and Governance Mechanisms

The management of green spaces in Kolkata involves multiple agencies such as KMC, KMDA, PWD, and the State Forest Department, leading to fragmentation and overlap in responsibilities. A unified institutional framework such as a City Green Infrastructure Management Authority (CGIMA) should be established to coordinate policy implementation, data management, and inter-departmental communication. This authority would be responsible for preparing periodic Urban Green Audits, setting sustainability targets, and ensuring compliance with environmental norms.

3. Ensuring Equitable Access to Urban Green Spaces

Equitable access to green spaces must be recognized as a basic urban right. The uneven distribution of green areas across Kolkata's 16 boroughs reveals socio-spatial disparities, where densely populated and economically weaker zones often lack adequate open spaces. Policy measures should aim to ensure that every citizen has access to a green space within a 500-m radius, as recommended by the World Health Organization (WHO). Urban redevelopment projects must include mandatory green components, while underutilized public lands, rooftops, and vacant plots should be converted into micro-parks, vertical gardens, and community green areas.

4. Promoting Public–Private and Community Partnerships

Given the financial and administrative constraints faced by local bodies, public–private partnerships (PPPs) and community-based management models are crucial for sustainable green space maintenance. Corporate entities can adopt parks under Corporate Social Responsibility (CSR) initiatives, contributing to their upkeep and biodiversity enhancement. Moreover, local communities and resident welfare associations should be actively engaged in participatory planning, tree plantation drives, and environmental monitoring, fostering a sense of stewardship and ownership over urban green assets.

5. Conservation of Biodiversity and Ecological Assets

Kolkata's wetlands, riverbanks, and urban forests such as the East Kolkata Wetlands and Maidan represent critical ecological resources that require priority conservation. Strict enforcement of environmental regulations and the implementation of ecological zoning can prevent further degradation of these assets. Policies should promote native species plantation, biodiversity parks, and urban wildlife corridors to enhance ecological connectivity. Additionally, ecosystem-based adaptation (EbA) strategies should be adopted to address urban climate resilience through nature-based solutions.

6. Policy Innovation and Long-Term Monitoring

Sustainability requires continuous policy evaluation and adaptive management. Establishing an Urban Green Observatory under KMC's jurisdiction can facilitate real-time data collection, satellite monitoring, and public reporting on green space status. Periodic reviews of environmental indicators such as per capita green cover, tree canopy density, and land surface temperature will help assess the effectiveness of implemented measures.

7. Integrating Health and Climate Goals with Green Policy

Urban green planning should be aligned with broader national initiatives such as the National Action Plan on Climate Change (NAPCC), Smart Cities Mission, and National Urban Policy Framework (NUPF). Recognizing UGS as natural infrastructure that supports public health, climate mitigation, and social well-being will strengthen their role in achieving Sustainable Development Goals (SDGs 3, 11, and 13).

In conclusion, sustainable urban planning in Kolkata demands a paradigm shift from viewing green spaces as ornamental to recognizing them as essential ecological and social infrastructure. By integrating robust governance, equitable access, community participation, and scientific management, the city can ensure the preservation and expansion of its urban ecological assets, ultimately fostering a greener, healthier, and more resilient urban future.

7.6 Formulation of a Consolidated Policy Framework Specifically Tailored to Strengthen the Governance, Planning and Management of Urban Green Spaces in the Context of Environmental Sustainability and Climate Resilience

The formulation of a consolidated policy framework for Urban Green Spaces (UGS) is imperative for fostering environmentally sustainable and climate-resilient cities. In the context of the Kolkata Municipal Corporation (KMC), where urban expansion and population pressure have significantly reduced the extent and quality of green cover, a comprehensive, multi-dimensional policy framework is essential to safeguard ecological assets, enhance environmental governance, and ensure equitable access to urban greenery. This policy framework must integrate environmental, social, and institutional dimensions, aligning with national and international sustainability agendas to achieve long-term resilience and livability.

1. Policy Vision and Objectives

The primary vision of this framework should be to create a resilient and inclusive green urban ecosystem that balances development with ecological preservation. The objectives include:

- Strengthening institutional governance and inter-agency coordination for green space management.
- Integrating UGS into urban planning and development processes.
- Enhancing climate resilience through nature-based solutions.
- Promoting equitable access to green spaces for all socio-economic groups.
- Encouraging participatory management and public–private partnerships for sustainable maintenance.

2. Institutional Strengthening and Governance Mechanisms

A major shortcoming in Kolkata's existing urban governance structure lies in the fragmentation of responsibilities among multiple agencies, including KMC, KMDA, the State Forest Department, and PWD. To overcome this, the policy framework should establish a Unified Green Governance Authority (UGGA). This centralized body would act as a nodal agency for planning, regulating, and monitoring UGS across the city. The UGGA should be empowered to coordinate across departments, ensure compliance with land use norms, and implement green zoning regulations. It should also facilitate inter-institutional collaboration with academic institutions, NGOs, and citizen groups to strengthen data sharing and policy innovation.

3. Integration of Green Infrastructure into Urban Planning

The framework should institutionalize the integration of Urban Green Infrastructure (UGI) into city master plans and development projects. Green spaces must be recognized as essential infrastructure, similar to transport or housing, with dedicated budget allocations and legal protection. The policy should enforce the Urban and Regional Development Plans Formulation and Implementation (URDPFI) Guidelines, 2014, which recommend allocating at least 15–20% of city land for green and open spaces. Spatial planning tools like GIS-based green audits, remote sensing, and urban ecological mapping should be used to identify green-deficient zones and monitor landscape transformation.

4. Legal Protection and Financial Mechanisms

The policy framework must include robust legal instruments to prevent the encroachment and conversion of green spaces. Mandatory Green Impact Assessments (GIA) should be introduced for major infrastructure and real estate projects, ensuring that ecological impacts are evaluated prior to approval. Financial sustainability can be achieved through dedicated green funds, generated via environmental taxes, developer contributions, and Corporate Social Responsibility (CSR) investments. Incentivizing green initiatives, such as vertical gardens, rooftop greenery, and biodiversity parks, can further mobilize resources for urban ecological enhancement.

5. Community Participation and Public–Private Partnerships

Sustainable management of UGS requires active involvement of citizens and private stakeholders. The policy should establish co-management models that empower Resident Welfare Associations (RWAs), NGOs, and corporate entities to participate in park maintenance, tree planting, and biodiversity conservation. Public awareness programs, green education campaigns, and community-driven initiatives like "Adopt-a-Park" schemes can foster environmental stewardship and local ownership of green assets.

6. Enhancing Climate Resilience through Urban Green Strategies

Urban green spaces are integral to climate resilience. The framework should prioritize nature-based solutions (NbS) such as green corridors, urban wetlands restoration, and tree canopy expansion to mitigate the urban heat island (UHI) effect and reduce carbon emissions. It should align with national initiatives like the National Action Plan on Climate Change (NAPCC) and the State Action Plan on Climate Change (SAPCC) to strengthen climate adaptation strategies through urban greenery.

7. Monitoring, Evaluation, and Policy Accountability

A robust monitoring and evaluation mechanism is crucial for ensuring the accountability and success of this framework. The establishment of a City Green Observatory (CGO) under the UGGA can facilitate periodic data collection, satellite-based mapping, and public disclosure of the state of urban green spaces. Key performance indicators such as per capita green cover, biodiversity index, and land surface temperature reduction should be used to assess progress and policy impact.

In conclusion, the proposed consolidated policy framework for UGS governance envisions an integrated, participatory, and technologically enabled approach to urban ecological management. By uniting governance reforms, spatial planning, legal enforcement, and community engagement, this framework can transform Kolkata into a model of green urban resilience. Prioritizing UGS as critical natural infrastructure will not only advance environmental sustainability but also improve urban livability, mitigate climate risks, and secure the ecological future of the city for generations to come.

Photo Plates

Some Insights During Field Survey

© The Editor(s) (if applicable) and The Author(s), under exclusive license
to Springer Nature Switzerland AG 2026
M. B. Ali and S. Jamal, *Urban Green Spaces and Environmental Sustainability*,
https://doi.org/10.1007/978-3-032-15322-7

MI 11 LITE | BABAR

MI 11 LITE | BABAR

Glossary

Amalthas Golden shower tree (A tropical deciduous tree)
Awastha Sacred Fig (A tropical deciduous tree)
Bagan Garden
Bagicha Garden
Bakul Mimusops elengi (A tropical deciduous tree)
Banyan Ficus benghalensis (A tropical deciduous tree)
Borough Sub-division of the municipal corporation and it is bigger than municipal ward
EKTARA A local NGO in Kolkata Municipal Corporation
Hijal Indian Oak Tree
Jam Indian blackberry (A tropical deciduous tree)
Jarul Lagerstroemia speciosa is a species of Lagerstroemia native to tropical southern Asia
Karanja Millettia pinnata (A tropical deciduous tree)
Maidan A field with trees and grasses
Math Field
Neem Azadirachta indica A (A tropical deciduous tree)
Peepal Ficus religiosa (A tropical deciduous tree)
Pukur Pond
Sirish Indian siris (A tropical deciduous tree)
Sita Asok Saraca asoca (A tropical deciduous tree)
Talab Pond
Uddyan Park

© The Editor(s) (if applicable) and The Author(s), under exclusive license to Springer Nature Switzerland AG 2026

M. B. Ali and S. Jamal, *Urban Green Spaces and Environmental Sustainability*,
https://doi.org/10.1007/978-3-032-15322-7

Bibliography

Abass, K., Afriyie, K., & Gyasi, R. M. (2019). From green to grey: The dynamics of land use/land cover change in urban Ghana. *Landscape Research, 44*(8), 909–921.

Aguilera, I., Foraster, M., Basagaña, X., Corradi, E., Deltell, A., Morelli, X., … & Künzli, N. (2015). Application of land use regression modelling to assess the spatial distribution of road traffic noise in three European cities. *Journal of Exposure Science & Environmental Epidemiology, 25*(1), 97–105.

Ajmal, U., Jamal, S., Ahmad, W. S., Ali, M. A., & Ali, M. B. (2022). Waterborne diseases vulnerability analysis using fuzzy analytic hierarchy process: A case study of Azamgarh city, India. *Modeling Earth Systems and Environment, 8*(2), 2687–2713.

Aklibasinda, M., & Ozdarici Ok, A. (2019). Determination of the urbanization and changes in open-green spaces in Nevsehir city through remote sensing. *Environmental Monitoring and Assessment, 191*(12), 756.

Alavipanah, S., Wegmann, M., Qureshi, S., Weng, Q., & Koellner, T. (2015). The role of vegetation in mitigating urban land surface temperatures: A case study of Munich, Germany during the Warm Season. *Sustainability, 7*(4), 4689–4706.

Ali, M. U., Liu, G., Yousaf, B., Ullah, H., Abbas, Q., & Munir, M. A. M. (2019). A systematic review on global pollution status of particulate matter-associated potential toxic elements and health perspectives in urban environment. *Environmental Geochemistry and Health, 41*, 1131–1162.

Ali, S. A., & Ahmad, A. (2018). Using analytic hierarchy process with GIS for Dengue risk mapping in Kolkata Municipal Corporation, West Bengal, India. *Spatial Information Research, 26*(4), 449–469.

Allen, J., & Lu, K. (2003). Modeling and prediction of future urban growth in the Charleston region of South Carolina: A GIS-based integrated approach. *Conservation Ecology, 8*(2).

Al-shalabi, M., Billa, L., Pradhan, B., Mansor, S., & Al-Sharif, A. A. (2013). Modelling urban growth evolution and land-use changes using GIS based cellular automata and SLEUTH models: The case of Sana'a metropolitan city, Yemen. *Environmental Earth Sciences, 70*(1), 425–437.

Al-sharif, A. A., & Pradhan, B. (2015). A novel approach for predicting the spatial patterns of urban expansion by combining the chi-squared automatic integration detection decision tree, Markov chain and cellular automata models in GIS. *Geocarto International, 30*(8), 858–881.

Altunkasa, M. F., & Uslu, C. (2004). The effects of urban green spaces on house prices in the upper northwest urban development area of Adana (Turkey). *Turkish Journal of Agriculture and Forestry, 28*(3), 203–209.

Annerstedt Van Den Bosch, M., Mudu, P., Uscila, V., Barrdahl, M., Kulinkina, A., Staatsen, B., … & Egorov, A. I. (2016). Development of an urban green space indicator and the public health rationale. *Scandinavian Journal of Public Health, 44*(2), 159–167.

© The Editor(s) (if applicable) and The Author(s), under exclusive license to Springer Nature Switzerland AG 2026

M. B. Ali and S. Jamal, *Urban Green Spaces and Environmental Sustainability*, https://doi.org/10.1007/978-3-032-15322-7

Annerstedt, M., Östergren, P. O., Björk, J., Grahn, P., Skärbäck, E., & Währborg, P. (2012). Green qualities in the neighbourhood and mental health–results from a longitudinal cohort study in Southern Sweden. *BMC Public Health, 12*, 1–13.

Aram, F., Solgi, E., & Holden, G. (2019). The role of green spaces in increasing social interactions in neighborhoods with periodic markets. *Habitat International, 84*, 24–32.

Arantes, B. L. D., & Mauad, T. (2019). Urban forests, air quality and health: A systematic review. *International Forestry Review, 21*(2), 167–181.

Ascenso, A., Augusto, B., Silveira, C., Rafael, S., Coelho, S., Monteiro, A., ... & Miranda, A. I. (2021). Impacts of nature-based solutions on the urban atmospheric environment: A case study for Eindhoven, The Netherlands. *Urban Forestry & Urban Greening, 57*, 126870.

Asgarian, A., Amiri, B. J., & Sakieh, Y. (2015). Assessing the effect of green cover spatial patterns on urban land surface temperature using landscape metrics approach. *Urban Ecosystem, 18*(1), 209–222. https://doi.org/10.1007/s11252-014-0387-7

Askne, J., Santoro, M., Smith, G., & Fransson, J. E. (2003). Multitemporal repeat-pass SAR interferometry of boreal forests. *IEEE Transactions on Geoscience and Remote Sensing, 41*(7), 1540–1550.

Asta, F., Michelozzi, P., Cesaroni, G., De Sario, M., Badaloni, C., Davoli, M., & Schifano, P. (2019). The modifying role of socioeconomic position and greenness on the short-term effect of heat and air pollution on preterm births in Rome, 2001–2013. *International Journal of Environmental Research and Public Health, 16*(14), 2497.

Astell-Burt, T., & Feng, X. (2019). Association of urban green space with mental health and general health among adults in Australia. *JAMA Network Open, 2*(7), e198209–e198209.

Awasthi, A., Chauhan, S. S., & Goyal, S. K. (2011). A multi-criteria decision making approach for location planning for urban distribution centers under uncertainty. *Mathematical and Computer Modelling, 53*(1–2), 98–109.

Tuzin, B., Leeuwen, E., Rodenburg, C., & Peter, N. (2002). *Paper presented at the 38th International Planning Congress on "The Pulsar Effect" Planning with Peaks, Glifada, Athens*, 21–26 September 2002.

Badarinath, K. V. S., Kiran Chand, T. R., Madhavilatha, K., & Raghavaswamy, V. (2005). Studies on urban heat islands using envisat AATSR data. *Journal of the Indian Society of Remote Sensing, 33*(4), 495–501.

Badiu, D. L., Iojă, C. I., Pătroescu, M., Breuste, J., Artmann, M., Niță, M. R., ... & Onose, D. A. (2016). Is urban green space per capita a valuable target to achieve cities' sustainability goals? Romania as a case study. *Ecological Indicators, 70*, 53–66.

Balram, S., & Dragićević, S. (2005). Attitudes toward urban green spaces: Integrating questionnaire survey and collaborative GIS techniques to improve attitude measurements. *Landscape and Urban Planning, 71*(2–4), 147–162.

Banti, M. A, Kiachidis, K., & Gemitzi, A. (2019). Estimation of spatio-temporal vegetation trends in different land use environments across Greece. *Journal of Land Use Science, 14*(1), 21–36.

Bardhan, R., Debnath, R., & Bandopadhyay, S. (2016a). A conceptual model for identifying the risk susceptibility of urban green spaces using geo-spatial techniques. *Modeling Earth Systems and Environment, 2*(3), 1–12.

Bardhan, R., Debnath, R., & Bandopadhyay, S. (2016b). A conceptual model for identifying the risk susceptibility of urban green spaces using geo-spatial techniques. *Modeling Earth Systems and Environment, 2*, 1–12.

Baró, F., & Gómez-Baggethun, E. (2017). Assessing the potential of regulating ecosystem services as nature-based solutions in urban areas. In *Nature-based solutions to climate change adaptation in urban areas: Linkages between science, policy and practice* (pp. 139–158).

Bauduceau, N., Berry, P., Cecchi, C., Elmqvist, T., Fernandez, M., Hartig, T., ... & Tack, J. (2015). *Towards an EU research and innovation policy agenda for nature-based solutions & re-naturing cities: Final report of the horizon 2020 expert group on nature-based solutions and re-naturing cities.*

Becker, F., & Li, Z.-L. (1990). Toward a local split-window method over land surface. *International Journal of Remote Sensing, 11*(3), 369–393. https://doi.org/10.1080/01431169008955028

Bhagat, R. B., & Mohanty, S. (2009). Emerging pattern of urbanization and the contribution of migration in urban growth in India. *Asian Population Studies, 5*(1), 5–20.

Bhatta, B. (2009). Analysis of urban growth pattern using remote sensing and GIS: A case study of Kolkata, India. *International Journal of Remote Sensing, 30*(18), 4733–4746.

Bhattacharya, T., Mukhopadhyay, S., & Bhattacharya, D. (2022). Urban green spaces: Potentials and challenges of vertical gardens in Kolkata. *Urban Forestry & Urban Greening, 66*, 127427.

Bindi, M., Brandani, G., Dessì, A., Dibari, C., Ferrise, R., Moriondo, M., & Trombi, G. (2009). Impact of climate change on agricultural and natural ecosystems. *American Journal of Environmental Sciences, 5*(5), 633–638.

Biswas, M., Banerji, S., & Mitra, D. (2019). Land-use–land-cover change detection and application of Markov model: A case study of Eastern part of Kolkata. *Environment, Development and Sustainability*, 1–20.

Bolund, P., & Hunhammar, S. (1999). Ecosystem services in urban areas. *Ecological Economics, 29*(2), 293–301.

Bowler, D. E., Buyung-Ali, L., Knight, T. M., & Pullin, A. S. (2010). Urban greening to cool towns and cities: A systematic review of the empirical evidence. *Landscape and Urban Planning, 97*(3), 147–155.

Brambilla, G., & Gallo, V. (2016). QUIETE: A scheme for a new index of the environmental quality of green areas. *Noise Mapping, 3*(1).

Brambilla, G., & Maffei, L. (2006). Responses to noise in urban parks and in rural quiet areas. *Acta Acustica United with Acustica, 92*(6), 881–886.

Brambilla, G., Gallo, V., Asdrubali, F., & D'Alessandro, F. (2013). The perceived quality of soundscape in three urban parks in Rome. *The Journal of the Acoustical Society of America, 134*(1), 832–839.

Bryant, M. M. (2006). Urban landscape conservation and the role of ecological greenways at local and metropolitan scales. *Landscape and Urban Planning, 76*(1–4), 23–44.

Burns, J., Boogaard, H., Polus, S., Pfadenhauer, L. M., Rohwer, A. C., van Erp, A. M., ... & Rehfuess, E. (2019). Interventions to reduce ambient particulate matter air pollution and their effect on health. *Cochrane Database of Systematic Reviews*, (5).

Buyadi, S., Mohd, W., & Misni, A. (2013). Impact of land use changes on the surface temperature distribution of area surrounding the National Botanic Garden, Shah Alam. *Procedia - Social and Behavioral Sciences, 101*, 516–525.

Byomkesh, T., Nakagoshi, N., & Dewan, A. M. (2012). Urbanization and green space dynamics in Greater Dhaka, Bangladesh. *Landscape and Ecological Engineering, 8*, 45–58.

Byrne, J. (2012). When green is White: The cultural politics of race, nature and social exclusion in a Los Angeles urban national park. *Geoforum, 43*(3), 595–611.

Byrne, J., & Wolch, J. (2009). Nature, race, and parks: Past research and future directions for geographic research. *Progress in Human Geography, 33*(6), 743–765.

Callaghan, A., McCombe, G., Harrold, A., McMeel, C., Mills, G., Moore-Cherry, N., & Cullen, W. (2021). The impact of green spaces on mental health in urban settings: A scoping review. *Journal of Mental Health, 30*(2), 179–193.

Canetti, A., Garrastazu, M. C., de Mattos, P. P., Braz, E. M., & Netto, S. P. (2018). Understanding multi-temporal urban forest cover using high resolution images. *Urban Forestry & Urban Greening, 29*, 106–112.

Carrus, G., Scopelliti, M., Lafortezza, R., Colangelo, G., et al. (2015a). Go greener, feel better? The positive efects of biodiversity on the well-being of individuals visiting urban and peri-urban green areas. *Landscape and Urban Planning, 134*, 221–228. https://doi.org/10.1016/j.landurbplan.2014.10.022

Carrus, G., Scopelliti, M., Lafortezza, R., Colangelo, G., Ferrini, F., Salbitano, F., … & Sanesi, G. (2015b). Go greener, feel better? The positive effects of biodiversity on the well-being of individuals visiting urban and peri-urban green areas. *Landscape and Urban Planning, 134*, 221–228.

Castle, C. J., & Crooks, A. T. (2006). Principles and concepts of agent-based modelling for developing geospatial simulations. Centre for Contemporary Communication Kolkata. www.ccckolkata.org

Census. (2011). https://censusindia.gov.in/census.website/

Chadchan, J., & Shankar, R. (2012). An analysis of urban growth trends in the post-economic reforms period in India. *International Journal of Sustainable Built Environment, 1*(1), 36–49.

Chakraborti, S., Das, D. N., Mondal, B., Shafizadeh-Moghadam, H., & Feng, Y. (2018). A neural network and landscape metrics to propose a flexible urban growth boundary: A case study. *Ecological Indicators, 93*, 952–965.

Chakraborty, A., Sengupta, S., & Roy, M. (2020). Urban green spaces and mental well-being: An empirical study of Kolkata City, India. *Habitat International, 104*, 102278.

Chandan, M. C., Nimish, G., & Bharath, H. A. (2020). Analysing spatial patterns and trend of future urban expansion using SLEUTH. *Spatial Information Research, 28*(1), 11–23.

Chen, X. L., Zhao, H. M., Li, P. X., & Yin, Z. Y. (2006a). Remote sensing image-based analysis of the relationship between urban heat island and land use/cover changes. *Remote Sensing of Environment, 104*(2), 133–146.

Chen, Y. H., Wang, J., & Li, X. B. (2002). A study on urban thermal field in summer based on satellite remote sensing. *Remote Sensing for Land Resources, 4*(1).

Chen, J., Guo, F., & Wu, Y. (2011). One decade of urban housing reform in China: Urban housing price dynamics and the role of migration and urbanization, 1995–2005. *Habitat International, 35*(1), 1–8.

Chen, J., Zhou, C., Wang, S., & Hu, J. (2018). Identifying the socioeconomic determinants of population exposure to particulate matter (PM2. 5) in China using geographically weighted regression modeling. *Environmental Pollution, 241*, 494–503.

Chen, W. Y. (2015). The role of urban green infrastructure in offsetting carbon emissions in 35 major Chinese cities: A nationwide estimate. *Cities, 44*, 112–120.

Chen, X. L., Zhao, H. M., Li, P. X., & Yin, Z. Y. (2006b). Remote sensing image-based analysis of the relationship between urban heat island and land use/cover changes. *Remote Sensing of Environment, 104*(2), 133–146.

Chen, X., Zhang, L. W., Huang, J. J., Song, F. J., Zhang, L. P., Qian, Z. M., … & Tang, N. J. (2016). Long-term exposure to urban air pollution and lung cancer mortality: A 12-year cohort study in Northern China. *Science of the Total Environment, 571*, 855–861.

Chong, S., Lobb, E., Khan, R., Abu-Rayya, H., Byun, R., & Jalaludin, B. (2013). Neighbourhood safety and area deprivation modify the associations between parkland and psychological distress in Sydney, Australia. *BMC Public Health, 13*(1), 1–8.

Chudnovsky, A., Ben-Dor, E., & Saaroni, H. (2004). Diurnal thermal behavior of selected urban objects using remote sensing measurements. *Energy and Buildings, 36*, 1063–1074. https://doi.org/10.1016/j.enbuild.2004.01.052

Cicea, C., & Pirlogea, C. (2011). Green spaces and public health in urban areas. *Theoretical and Empirical Researches in Urban Management, 6*(1), 83–92.

Clarke, M., Wallace, C., Cadaval, S., Anderson, E., Egerer, M., Dinkins, L., & Platero, R. (2023). Factors that enhance or hinder social cohesion in urban greenspaces: A literature review. *Urban Forestry & Urban Greening*, 127936.

Cohen, A. J., Brauer, M., Burnett, R., Anderson, H. R., Frostad, J., Estep, K., … & Forouzanfar, M. H. (2017). Estimates and 25-year trends of the global burden of disease attributable to ambient air pollution: An analysis of data from the Global Burden of Diseases Study 2015. *The Lancet, 389*(10082), 1907–1918.

Cohen, B. (2004). Urban growth in developing countries: A review of current trends and a caution regarding existing forecasts. *World Development, 32*(1), 23–51.

Cohen, B. (2006). Urbanization in developing countries: Current trends, future projections, and key challenges for sustainability. *Technology in Society, 28*(1–2), 63–80.

Cohen, J. (1960). A coefficient of agreement for nominal scales. *Educational and Psychological Measurement, 20*(1), 37–46.

Connop, S., Vandergert, P., Eisenberg, B., Collier, M. J., Nash, C., Clough, J., & Newport, D. (2016). Renaturing cities using a regionally-focused biodiversity-led multifunctional benefits approach to urban green infrastructure. *Environmental Science & Policy, 62*, 99–111.

CPCB. (2020). https://cpcb.nic.in/

Dahal, K. R., Benner, S., & Lindquist, E. (2018). Analyzing spatiotemporal patterns of urbanization in Treasure Valley, Idaho, USA. *Applied Spatial Analysis and Policy, 11*(2), 205–226.

Dallimer, M., Tang, Z., Bibby, P. R., Brindley, P., Gaston, K. J., & Davies, Z. G. (2011). Temporal changes in greenspace in a highly urbanized region. *Biology Letters, 7*(5), 763–766.

Dallimer, M., Tang, Z., Gaston, K. J., & Davies, Z. G. (2016). The extent of shifts in vegetation phenology between rural and urban areas within a human-dominated region. *Ecology and Evolution, 6*(7), 1942–1953.

Das, S., Ghosh, S., Chakraborty, P., & Mondal, S. (2021). Urban green spaces and community participation: An analysis of Kolkata City. *Cities, 117*, 103307.

Demographia World Urban Areas 13th Annual Edition. (2017). https://www.demographia.com/dbworldua-index.htm

Deng, C., & Wu, C. (2012). BCI: A biophysical composition index for remote sensing of urban environments. *Remote Sensing of Environment, 127*, 247–259. https://doi.org/10.1016/j.rse.2012.09.009

Dewan, A. M., & Corner, R. J. (2014). Impact of land use and land cover changes on urban land surface temperature. *Dhaka Megacity: Geospatial Perspectives on Urbanisation, Environment and Health*, 219–238.

Dewan, A. M., & Yamaguchi, Y. (2009). Land use and land cover change in Greater Dhaka, Bangladesh: Using remote sensing to promote sustainable urbanization. *Applied Geography, 29*(3), 390–401.

Di, Q., Wang, Y., Zanobetti, A., Wang, Y., Koutrakis, P., Choirat, C., … & Schwartz, J. D. (2017). Air pollution and mortality in the Medicare population. *New England Journal of Medicine, 376*(26), 2513–2522.

Diener, A., & Mudu, P. (2021). How can vegetation protect us from air pollution? A critical review on green spaces' mitigation abilities for air-borne particles from a public health perspective-with implications for urban planning. *Science of the Total Environment, 796*, 148605.

Dinda, S., Chatterjee, N. D., & Ghosh, S. (2021). An integrated simulation approach to the assessment of urban growth pattern and loss in urban green space in Kolkata, India: A GIS-based analysis. *Ecological Indicators, 121*, 107178.

Dinda, S., Chatterjee, N. D., & Ghosh, S. (2022). Modelling the future vulnerability of urban green space for priority-based management and green prosperity strategy planning in Kolkata, India: A PSR-based analysis using AHP-FCE and ANN-Markov model. *Geocarto International, 37*(22), 6551–6578.

DoUD&MA. (2016). *The Green City mission*. Department of Urban Development & Municipal Affairs (DoUD&MA), Government of West Bengal. Accessed on 2nd May, 2020.

Douglas, O., Lennon, M., & Scott, M. (2017). Green space benefits for health and well-being: A life-course approach for urban planning, design and management. *Cities, 66*, 53–62.

Doygun, H., & İlter, A. A. (2007). Investigation of current and projected active green area adequacy in Kahramanmaraş City. *Journal of Ecology, 16*(65).

Dutta, S., Sahana, M., & Guchhait, S. K. (2017). Assessing anthropogenic disturbance on forest health based on fragment grading in Durgapur Forest Range, West Bengal, India. *Spatial Information Research, 25*(3), 501–512.

Dyson, B. (2017). *Integration of life cycle assessment into decision-analytic approaches for sustainable technologies.*

Dzhambov, A. M., & Dimitrova, D. D. (2014). Urban green spaces' effectiveness as a psychological buffer for the negative health impact of noise pollution: A systematic review. *Noise and Health, 16*(70), 157.

Eastman, J. R. (2016). *TerrSet geospatial monitoring and modelling system.* Clark Labs, Clark University.

Egorov, A. I., Mudu, P., Braubach, M., & Martuzzi, M. (2016). *Urban green spaces and health—A review of evidence.* WHO Regional Office for Europe.

EKTARA. https://ektara.org.in

Enssle, F., & Kabisch, N. (2020). Urban green spaces for the social interaction, health and well-being of older people—An integrated view of urban ecosystem services and socio-environmental justice. *Environmental Science & Policy, 109*, 36–44.

Ersoy, M. (2015). Kentsel planlamada standartlar. *İstanbul: Ninova Yayıncılık.*

Escobedo, F. J., Kroeger, T., & Wagner, J. E. (2011). Urban forests and pollution mitigation: Analyzing ecosystem services and disservices. *Environmental Pollution, 159*(8–9), 2078–2087. https://doi.org/10.1016/j.envpol.2011.01.010

Etzioni, A. (1998). *The essential communitarian reader*. Rowman and Butterfield.

European Commission Communication from the Commission to the European Parliament, the Council, the European Economic and Social Committee and the Committee of the Regions. Green Infrastructure (GI)–Enhancing Europe's Natural Capital; Brussels. Available online: https://eur-lex.europa.eu/resource.html?uri=cellar:d41348f2-01d5-4abe-b817-4c73e6f1b2df.0014.03/DOC_1&format=PDF

Fan, X., Liu, J., Chen, J., Zhao, L., Hong, W., Hong, T., & Lin, H. (2018). Effects of landscape changes on ecosystem services: A case study in Nanping, northern Fujian province, China. *Fresenius Environmental Bulletin, 27*(5A), 3561–3570.

Feng, H., Liu, H., & Wu, L. (2014). Monitoring the relationship between the land surface temperature change and urban growth in Beijing China. *IEEE Journal of Selected Topics in Applied Earth Observations and Remote Sensing, 7*(10), 4010–4019.

Ferreira, L. S., & Duarte, D. H. S. (2019). Exploring the relationship between urban form, land surface temperature and vegetation indices in a subtropical megacity. *Urban Climate, 27*, 105–123.

Foody, G. M. (1992). On the compensation for chance agreement in image classification accuracy assessment. *Photogrammetric Engineering and Remote Sensing, 58*(10), 1459–1460.

Forster, B. (1983). Some urban measurements from Landsat data. *Photogrammetric Engineering and Remote Sensing, 49*, 1693–1707.

Fuller, R. A., & Gaston, K. J. (2009). The scaling of green space coverage in European cities. *Biology Letters, 5*(3), 352–355.

Gairola, S., & Noresah, M. S. (2010). Emerging trend of urban green space research and the implications for safeguarding biodiversity: A viewpoint. *Nature and Science, 8*(7), 43–49. S

Ganaie, T. A., Jamal, S., & Ahmad, W. S. (2020). Changing land use/land cover patterns and growing human population in Wular catchment of Kashmir Valley, India. *GeoJournal*, 1–18.

Ganaie, T. A., Jamal, S., & Ahmad, W. S. (2021). Changing land use/land cover patterns and growing human population in Wular catchment of Kashmir Valley, India. *GeoJournal, 86*, 1589–1606.

Gao, B.-C. (1996). NDWI—A normalized difference water index for remote sensing of vegetation liquid water from space. *Remote Sensing of Environment, 58*(3), 257–266.

Gao, P., Niu, X., Wang, B., & Zheng, Y. (2015). Land use changes and its driving forces in hilly ecological restoration area based on GIS and RS of northern china. *Scientific Reports, 5*(1), 1–11.

Gavrilidis, A. A., Niță, M. R., Onose, D. A., Badiu, D. L., & Năstase, I. I. (2019). Methodological framework for urban sprawl control through sustainable planning of urban green infrastructure. *Ecological Indicators, 96*, 67–78.

Gazette, O. (2017). *Planned areas zoning regulation.* http://www.resmigazete.gov.tr/eskiler/2017/07/20170703-8.htm

Geng, D., Innes, J., Wu, W., & Wang, G. (2021). Impacts of COVID-19 pandemic on urban park visitation: A global analysis. *Journal of Forestry Research, 32*, 553–567.

Ghosh, S., Chatterjee, N. D., & Dinda, S. (2021). Urban ecological security assessment and forecasting using integrated DEMATEL-ANP and CA-Markov models: A case study on Kolkata Metropolitan Area, India. *Sustainable Cities and Society, 68*, 102773.

Ghosh, S., Dinda, S., Chatterjee, N. D., & Das, K. (2018). Analyzing risk factors for shrinkage and transformation of East Kolkata Wetland, India. *Spatial Information Research, 26*(6), 661–677.

Gibson, S., Loukaitou-Sideris, A., & Mukhija, V. (2019). Ensuring park equity: A California case study. *Journal of Urban Design, 24*(3), 385–405.

Gómez-Baggethun, E., & Barton, D. N. (2013). Classifying and valuing ecosystem services for urban planning. *Ecological Economics, 86*, 235–245.

Gonzalez, M. T., & Kirkevold, M. (2016). Design characteristics of sensory gardens in Norwegian nursing homes: A cross-sectional e-mail survey. *Journal of Housing for the Elderly, 30*(2), 141–155.

González-Oreja, J. A., Bonache-Regidor, C., & de la Fuente-Díaz, A. A. (2010). Far from the noisy world? Modelling the relationships between park size, tree cover and noise levels in urban green spaces of the city of Puebla, Mexico. *Interciencia, 35*(7), 486–492.

Gozalo, G. R., Morillas, J. M. B., González, D. M., & Moraga, P. A. (2018). Relationships among satisfaction, noise perception, and use of urban green spaces. *Science of the Total Environment, 624*, 438–450.

Grahn, P., & Stigsdotter, U. A. (2003). Landscape planning and stress. *Urban Forestry & Urban Greening, 2*(1), 1–18.

Grigoraş, G., & Uriţescu, B. (2019). Land use/land cover changes dynamics and their effects on surface urban heat island in Bucharest, Romania. *International Journal of Applied Earth Observation and Geoinformation, 80*, 115–126.

Grimm, N. B., Faeth, S. H., Golubiewski, N. E., Redman, C. L., Wu, J., Bai, X., & Briggs, J. M. (2008). Global change and the ecology of cities. *Science, 319*(5864), 756–760.

Grote, R., Samson, R., Alonso, R., Amorim, J. H., Cariñanos, P., Churkina, G., … & Calfapietra, C. (2016). Functional traits of urban trees: Air pollution mitigation potential. *Frontiers in Ecology and the Environment, 14*(10), 543–550.

Guo, G., Wu, Z., Xiao, R., & Chen, Y. (2015). Landscape and urban planning impacts of urban biophysical composition on land surface temperature in urban heat island clusters. *Landscape and Urban Planning, 135*, 1–10. https://doi.org/10.1016/j.landurbplan.2014.11.007

Gupta, K., Kumar, P., Pathan, S. K., & Sharma, K. P. (2012). Urban Neighborhood Green Index—A measure of green spaces in urban areas. *Landscape and Urban Planning, 105*(3), 325–335.

Haack, B. N., & Rafter, A. (2006). Urban growth analysis and modeling in the Kathmandu Valley, Nepal. *Habitat International, 30*(4), 1056–1065.

Han, Y., Kang, W., & Song, Y. (2018). Mapping and quantifying variations in ecosystem services of urban green spaces: A test case of carbon sequestration at the district scale for Seoul, Korea (1975–2015). *International Review for Spatial Planning and Sustainable Development, 6*(3), 110–120.

Hang, H. T., & Rahman, A. (2018). Characterization of thermal environment over heterogeneous surface of National Capital Region (NCR), India using LANDSAT-8 sensor for regional planning studies. *Urban Climate*, 1–18. https://doi.org/10.1016/j.uclim.2018.01.001

Hang, H. T., & Rahman, A. (2018b). Characterization of thermal environment over heterogeneous surface of National Capital Region (NCR), India using LANDSAT-8 sensor for regional planning studies. *Urban Climate, 24*, 1–18. https://doi.org/10.1016/j.uclim.2018.01.001

Hao, P., Niu, Z., Zhan, Y., Wu, Y., Wang, L., & Liu, Y. (2016). Spatiotemporal changes of urban impervious surface area and land surface temperature in Beijing from 1990 to 2014. *Giscience & Remote Sensing, 53*(1), 63–84. https://doi.org/10.1080/15481603.2015.1095471

Hartig, T., Mitchell, R., De Vries, S., & Frumkin, H. (2014). Nature and health. *Annual Review of Public Health, 35*, 207–228.

Hartmann, D. L., Tank, A. M. G. K., & Rusticucci, M. (2013). IPCC 5th assessment report, climate change 2013: The physical science basis. In *Paper presented at the IPCC*.

Haruna, A. I., Oppong, R. A., & Marful, A. B. (2018). Exploring eco-aesthetics for urban green infrastructure development and building resilient cities: A theoretical overview. *Cogent Social Sciences, 4*(1), 1478492.

Huang, D., Lu, C. C., & Wang, G. (2009). Integrated management of urban green space: The case in Guangzhou China, in *45th ISOCARP Congress 2009.*

Hashem, N., & Balakrishnan, P. (2015). Change analysis of land use/land cover and modelling urban growth in Greater Doha, Qatar. *Annals of GIS, 21*(3), 233–247.

Heidt, V., & Neef, M. (2008). Benefits of urban green space for improving urban climate. In *Ecology, planning, and management of urban forests* (pp. 84–96). Springer.

Heo, S., & Bell, M. L. (2019). The influence of green space on the short-term effects of particulate matter on hospitalization in the US for 2000–2013. *Environmental Research, 174*, 61–68.

Hofmann, P., Strobl, J., & Nazarkulova, A. (2011). Mapping green spaces in Bishkek—How reliable can spatial analysis be? *Remote Sensing, 3*(6), 1088–1103.

Horoshenkov, K. V., Khan, A., & Benkreira, H. (2013). Acoustic properties of low growing plants. *The Journal of the Acoustical Society of America, 133*(5), 2554–2565.

http://www.countdown2010.net/2010/wp.../FS7Greenspace_small.PDF

Huang, C., Yang, J., Lu, H., Huang, H., & Yu, L. (2017). Green spaces as an indicator of urban health: Evaluating its changes in 28 mega-cities. *Remote Sensing, 9*(12), 1266.

Huddart, L. (1990). *The use of vegetation for traffic noise screening* (No. 38).

Ichimura, M. (2003, January). Urbanization, urban environment and land use: challenges and opportunities. In *Asia-Pacific forum for environment and development, expert meeting* (Vol. 23, pp. 1–14).

Jamal, S., & Ahmad, W. S. (2020). Assessing land use land cover dynamics of wetland ecosystems using Landsat satellite data. *SN Applied Sciences, 2*, 1–24.

Jamal, S., & Ali, M. B. (2023). Determining urban growth in response to land use dynamics using multilayer perceptron and Markov chain models in a metropolitan city: past and future. *Environment, Development and Sustainability* 1–27.

Jamal, S., Ahmad, W. S., Ajmal, U., Aaquib, M., Ashif Ali, M., Babor Ali, M., & Ahmed, S. (2022a). An integrated approach for determining the anthropogenic stress responsible for degradation of a Ramsar Site-Wular Lake in Kashmir, India. *Marine Geodesy, 45*(4), 407–434.

Jamal, S., Ali, M. B., Ali, M. A., & Ajmal, U. (2022b). Evaluation and distribution of urban green spaces in Kolkata Municipal Corporation: An approach to urban sustainability. *Towards sustainable natural resources: Monitoring and managing ecosystem biodiversity* (pp. 151–172). Springer International Publishing.

Jamal, S., Saqib, M., Ahmad, W. S., Ahmad, M., Ali, M. A., & Ali, M. B. (2023). Unraveling the complexities of land transformation and its impact on urban sustainability through land surface temperature analysis. *Applied Geomatics*, 1–23.

Jamil, M., Sahana, M., & Sajjad, H. (2018). Crop suitability analysis in the Bijnor District, UP, using geospatial tools and fuzzy analytical hierarchy process. *Agricultural Research, 7*(4), 506–522.

Janhäll, S. (2015). Review on urban vegetation and particle air pollution—Deposition and dispersion. *Atmospheric Environment, 105*, 130–137.

Jat, M. K., Garg, P. K., & Khare, D. (2008a). Modelling of urban growth using spatial analysis techniques: A case study of Ajmer city (India). *International Journal of Remote Sensing, 29*(2), 543–567.

Jat, M. K., Garg, P. K., & Khare, D. (2008b). Monitoring and modelling of urban sprawl using remote sensing and GIS techniques. *International Journal of Applied Earth Observation and Geoinformation, 10*(1), 26–43.

Jaung, W., Carrasco, L. R., Shaikh, S. F. E. A., Tan, P. Y., & Richards, D. R. (2020). Temperature and air pollution reductions by urban green spaces are highly valued in a tropical city-state. *Urban Forestry & Urban Greening, 55*, 126827.

Jenks, M. J., Burgess, M. J. R., Acioly, C., Allen, A., Barter, P. A., & Brand, P. (2000). *Compact cities: Sustainable urban forms for developing countries.* Taylor & Francis.

Jennings, V., & Bamkole, O. (2019). The relationship between social cohesion and urban green space: An avenue for health promotion. *International Journal of Environmental Research and Public Health, 16*(3), 452.

Jensen, J. R. (2005). Digital image processing: A remote sensing perspective. Prentice Hall.

Jia, B. Q. (2013). Driving factor analysis on the vegetation changes derived from the Landsat TM images in Beijing. *Acta Ecologica Sinica, 33*(5), 1654–1666.

Jim, C. Y., & Chen, W. Y. (2008). Assessing the ecosystem service of air pollutant removal by urban trees in Guangzhou (China). *Journal of Environmental Management, 88*(4), 665–676.

Jorgensen, A., & Gobster, P. H. (2010). Shades of green: Measuring the ecology of urban green space in the context of human health and well-being. *Nature and Culture, 5*(3), 338–363.

Joshi, N., Joshi, A., & Bist, B. (2020). Phytomonitoring and mitigation of air pollution by plants. *Sustainable Agriculture in the Era of Climate Change*, 113–142.

Kabisch, N., & Van Den Bosch, M. A. (2017). Urban green spaces and the potential for health improvement and environmental justice in a changing climate. *Nature-Based Solutions to Climate Change Adaptation in Urban Areas: Linkages Between Science, Policy and Practice*, 207–220.

Kabisch, N., Strohbach, M., Haase, D., & Kronenberg, J. (2016). Urban green space availability in European cities. *Ecological Indicators, 70*, 586–596.

Kabisch, N., van den Bosch, M., & Lafortezza, R. (2017). The health benefits of nature-based solutions to urbanization challenges for children and the elderly—A systematic review. *Environmental Research, 159*, 362–373.

Kaczynski, A. T., Potwarka, L. R., & Saelens, B. E. (2008). Association of park size, distance, and features with physical activity in neighborhood parks. *American Journal of Public Health, 98*(8), 1451–1456.

Kaplan, R., & Kaplan, S. (1989). *The experience of nature: A psychological perspective*. Cambridge University Press.

Kar, R., Reddy, G. O., Kumar, N., & Singh, S. K. (2018). Monitoring spatio-temporal dynamics of urban and peri-urban landscape using remote sensing and GIS—A case study from Central India. *The Egyptian Journal of Remote Sensing and Space Science, 21*(3), 401–411.

Karuppannan, S., Baharuddin, Z. M., Sivam, A., & Daniels, C. B. (2014). Urban green space and urban biodiversity: Kuala Lumpur, Malaysia. *Journal of Sustainable Development, 7*(1), 1. https://doi.org/10.5539/jsd.v7n1p1

Kaufmann, R. K., Zhou, L., Myneni, R. B., Tucker, C. J., Slayback, D., Shabanov, N. V., Pinzon, J. (2003). The effect of vegetation on surface temperature: A statistical analysis of NDVI and climate data. *Geophysical Research Letters, 30*(22).

Kloek, M. E., Buijs, A. E., Boersema, J. J., & Schouten, M. G. (2013). Crossing borders: Review of concepts and approaches in research on greenspace, immigration and society in northwest European countries. *Landscape Research, 38*(1), 117–140.

Kogan, P., Gale, T., Arenas, J. P., & Arias, C. (2021). Development and application of practical criteria for the recognition of potential Health Restoration Soundscapes (HeReS) in urban greenspaces. *Science of the Total Environment, 793*, 148541.

Kolkata Municipal Corporation (KMC) Office.

Kolkata Municipal Development Authority (KMDA).

Kondo, M. C., Fluehr, J. M., McKeon, T., & Branas, C. C. (2018). Urban green space and its impact on human health. *International Journal of Environmental Research and Public Health, 15*(3), 445.

Kong, F., & Nakagoshi, N. (2006). Spatial-temporal gradient analysis of urban green spaces in Jinan, China. *Landscape and Urban Planning, 78*(3), 147–164.

Konijnendijk, C. C., Annerstedt, M., Nielsen, A. B., & Maruthaveeran, S. (2013). Benefts of urban parks. A systematic review. In *A Report for IFPRA*. Copenhagen & Alnarp. https://www.forskningsdatabasen.dk/en/catalog/2389131353

Koohsari, M. J., Mavoa, S., Villanueva, K., Sugiyama, T., Badland, H., Kaczynski, A. T., … & Giles-Corti, B. (2015). Public open space, physical activity, urban design and public health: Concepts, methods and research agenda. *Health & Place, 33*, 75–82.

Koprowska, K., Łaszkiewicz, E., Kronenberg, J., & Marcińczak, S. (2018). Subjective perception of noise exposure in relation to urban green space availability. *Urban Forestry & Urban Greening, 31*, 93–102.

Kumar, K. S., Bhaskar, P. U., & Padma Kumari, K. (2012). Estimation of land surface temperature to study urban heat island effect using Landsat ETM + IMAGE. *International Journal of Engineering, Science and Technology, 4*(02), 771–778.

Kumari, B., Tayyab, M., Mallick, J., Khan, M. F., & Rahman, A. (2018). Satellite-driven land surface temperature (LST) using Landsat 5, 7 (TM/ETM+ SLC) and Landsat 8 (OLI/TIRS) data and its association with built-up and green cover over urban Delhi, India. *Remote Sensing in Earth Systems Sciences, 1*(3), 63–78.

Kundu, A., Mukherjee, A., Banerjee, A., & Chatterjee, R. (2019). Green spaces and air quality: A study on Kolkata municipal corporation area, India. *Environmental Monitoring and Assessment, 191*(4), 238.

Loures, L., Santos, R., & Thomas, P. (2007). Urban parks and sustainable development: The case study of Partimao city, Portugal. In *Conference on Energy, Environment, Ecosystem and Sustainable Development*, Agios Nikolaos, Greece.

Lachowycz, K., & Jones, A. P. (2013). Towards a better understanding of the relationship between greenspace and health: Development of a theoretical framework. *Landscape and Urban Planning, 118*, 62–69.

Lafortezza, R., Davies, C., Sanesi, G., & Konijnendijk, C. C. (2013). Green Infrastructure as a tool to support spatial planning in European urban regions. *iForest-Biogeosciences and Forestry, 6*(3), 102.

Laing, R., Miller, D., Davies, A. M., & Scott, S. (2006). Urban green space: the incorporation of environmental values in a decision support system. *Journal of Information Technology in Construction, 11.*

Langemeyer, J., Camps-Calvet, M., Calvet-Mir, L., Barthel, S., & Gómez-Baggethun, E. (2018). Stewardship of urban ecosystem services: Understanding the value (s) of urban gardens in Barcelona. *Landscape and Urban Planning, 170*, 79–89.

Lee, A. C. K., Jordan, H. C., & Horsley, J. (2015). Value of urban green spaces in promoting healthy living and wellbeing: prospects for planning. *Risk Management and Healthcare Policy*, 131–137.

Lelieveld, J., Evans, J. S., Fnais, M., Giannadaki, D., & Pozzer, A. (2015). The contribution of outdoor air pollution sources to premature mortality on a global scale. *Nature, 525*(7569), 367–371.

Li, C. H., & Yang, Z. F. (2004). Spatio-temporal changes of NDVI and their relations with precipitation and Run of in the Yellow River Basin. *Geographical Research, 23*, 753–759.

Li, G., Zhang, H., Hu, M., He, J., Yang, W., Zhao, H., … & Huang, F. (2022). Associations of combined exposures to ambient temperature, air pollution, and green space with hypertension in rural areas of Anhui Province, China: A cross-sectional study. *Environmental Research, 204*, 112370.

Li, P., Yan, R., Yu, S., Wang, S., Liu, W., & Bao, H. (2015). Reinstate regional transport of PM2. 5 as a major cause of severe haze in Beijing. In *Proceedings of the National Academy of Sciences, 112*(21), E2739–E2740.

Li, X., Chen, G., Liu, X., Liang, X., Wang, S., Chen, Y., … & Xu, X. (2017). A new global land-use and land-cover change product at a 1-km resolution for 2010 to 2100 based on human–environment interactions. *Annals of the American Association of Geographers, 107*(5), 1040–1059.

Li, X., Mitra, C., Marzen, L., & Yang, Q. (2016). Spatial and temporal patterns of wetland cover changes in East Kolkata Wetlands, India from 1972 to 2011. *International Journal of Applied Geospatial Research (IJAGR), 7*(2), 1–13.

Lillesand, T., Kiefer, R. W., & Chipman, J. (2015). *Remote sensing and image interpretation*. Wiley.
Lin, B., Meyers, J., & Barnett, G. (2015). Understanding the potential loss and inequities of green space distribution with urban densification. *Urban Forestry & Urban Greening, 14*(4), 952–958.
Ling, T. Y., Hung, W. K., Lin, C. T., & Lu, M. (2020). Dealing with green gentrification and vertical green-related urban well-being: A contextual-based design framework. *Sustainability, 12*(23), 10020.
Liou, Y. A., & Mulualem, G. M. (2019). Spatio–temporal assessment of drought in Ethiopia and the impact of recent intense droughts. *Remote Sensing, 11*(15), 1828.
Liu, H. L., & Shen, Y. S. (2014). The impact of green space changes on air pollution and microclimates: A case study of the Taipei metropolitan area. *Sustainability, 6*(12), 8827–8855.
Liu, Q., Gu, X., Deng, F., Mu, L., Baccarelli, A. A., Guo, X., & Wu, S. (2019a). Ambient particulate air pollution and circulating C-reactive protein level: A systematic review and meta-analysis. *International Journal of Hygiene and Environmental Health, 222*(5), 756–764.
Liu, Q., Yang, Y., Tian, H., Zhang, B., & Gu, L. (2014a). Assessment of human impacts on vegetation in built-up areas in China based on AVHRR, MODIS and DMSP_OLS nighttime light data, 1992–2010. *Chinese Geographical Science, 24*, 231–244.
Liu, W., Zhan, J., Zhao, F., Yan, H., Zhang, F., & Wei, X. (2019b). Impacts of urbanization-induced land-use changes on ecosystem services: A case study of the Pearl River Delta Metropolitan Region, China. *Ecological Indicators, 98*, 228–238.
Liu, Y., Fang, P., Bian, D., Zhang, H., & Wang, S. (2014b). Fuzzy comprehensive evaluation for the motion performance of autonomous underwater vehicles. *Ocean Engineering, 88*, 568–577.
Loehle, C., & Scafetta, N. (2011). Climate change attribution using empirical decomposition of climatic data. *Open Atmospheric Science Journal, 5*, 1–4. https://doi.org/10.2174/1874282301105010074
Lopez, B. (2020). Parks are critical urban infrastructure: Perception and use of urban green spaces in NYC during COVID-19. *Preprints*.
Lopez, G. A. P., & Souza, L. C. L. D. (2018). Urban green spaces and the influence on vehicular traffic noise control. *Ambiente Construído, 18*, 161–175.
Lu, Y., Wang, X., Xie, Y., Li, K., & Xu, Y. (2016). Integrating future land use scenarios to evaluate the spatio-temporal dynamics of landscape ecological security. *Sustainability, 8*(12), 1242.
Luo, L., & Mountrakis, G. (2010). Integrating intermediate inputs from partially classified images within a hybrid classification framework: An impervious surface estimation example. *Remote Sensing of Environment, 114*(6), 1220–1229.
Luy, L., Ma, A., & Ma, Q. (2012). Spatial and temporal variations of vegetation coverage in coastal peri-urban area: A case study of Laoshan District, Qingdao. *Environmental Science & Technology (China), 35*(1), 178–185.
Luz, A. C., Buijs, M., Aleixo, C., Metelo, I., Grilo, F., Branquinho, C., & Pinho, P. (2019). Should I stay or should I go? Modelling the fluxes of urban residents to visit green spaces. *Urban Forestry & Urban Greening, 40*, 195–203.
Ma, Y., Kuang, Y., & Huang, N. (2010). Coupling urbanization analyzes for studying urban thermal environment and its interplay with biophysical parameters based on TM/ETM+imagery. *International Journal of Applied Earth Observation and Geoinformation, 12*(2), 110–118. https://doi.org/10.1016/j.jag.2009.12.002
Maas, J., Spreeuwenberg, P., Van Winsum-Westra, M., Verheij, R. A., Vries, S., & Groenewegen, P. P. (2009). Is green space in the living environment associated with people's feelings of social safety? *Environment and Planning A, 41*(7), 1763–1777.
Maas, J., Verheij, R. A., Groenewegen, P. P., De Vries, S., & Spreeuwenberg, P. (2006). Green space, urbanity, and health: How strong is the relation? *Journal of Epidemiology & Community Health, 60*(7), 587–592.
Maiti, S., & Agrawal, P. K. (2005). Environmental degradation in the context of growing urbanization: A focus on the metropolitan cities of India. *Journal of Human Ecology, 17*(4), 277–287.

Mallick, J., Kant, Y., & Bharath, B. D. (2008). Estimation of land surface temperature over Delhi using Landsat-7 ETM. *Journal of Indian Geophysical Union, 12*(3), 131–140.

Mallick, J., Rahman, A., & Singh, C. K. (2013). Modeling urban heat islands in heterogeneous land surface and its correlation with impervious surface area by using nighttime ASTER satellite data in highly urbanizing city, Delhi-India. *Advances in Space Research, 52*, 639–655. https://doi.org/10.1016/j.asr.2013.04.025

Mandal, J., Ghosh, N., & Mukhopadhyay, A. (2019). Urban growth dynamics and changing land-use land-cover of megacity Kolkata and its environs. *Journal of the Indian Society of Remote Sensing, 47*(10), 1707–1725.

Mansour, S., Al-Belushi, M., & Al-Awadhi, T. (2020). Monitoring land use and land cover changes in the mountainous cities of Oman using GIS and CA-Markov modelling techniques. *Land Use Policy, 91*, 104414.

Marcus, C. C., & Francis, C. (Eds.). (1997). *People places: Design guidelines for urban open space.* Wiley.

Margaritis, E., & Kang, J. (2016). Relationship between urban green spaces and other features of urban morphology with traffic noise distribution. *Urban Forestry & Urban Greening, 15*, 174–185.

Margaritis, E., & Kang, J. (2017). Relationship between green space-related morphology and noise pollution. *Ecological Indicators, 72*, 921–933.

Markevych, I., Schoierer, J., Hartig, T., Chudnovsky, A., Hystad, P., Dzhambov, A. M., … & Fuertes, E. (2017). Exploring pathways linking greenspace to health: Theoretical and methodological guidance. *Environmental Research, 158*, 301–317.

Martinez-Gonzalez, M. A., Varo, J. J., Santos, J. L., Irala, J. D., Gibney, M. J., Kearney, J., & Martinez, J. A. (2001). Prevalence of physical activity during leisure time in the European Union.

Mass, J. F. (1999). Monitoring land-cover changes: A comparison of change detection techniques. *International Journal of Remote Sensing, 20*(1), 139–152.

Mathew, A., Sreekumar, S., Khandelwal, S., Kaul, N., & Kumar, R. (2016). Prediction of surface temperatures for the assessment of urban heat island effect over Ahmedabad city using linear time series model. *Energy and Buildings, 128*, 605–616. https://doi.org/10.1016/j.enbuild.2016.07.004

McConnell, V., & Walls, M. (2005). Assessing the non-market value of open space. In *Report, resources for the future.*

McDonald, R., Kroeger, T., Boucher, T., Wang, L., & Salem, R. (2016). Planting healthy air: A global analysis of the role of urban trees in addressing particulate matter pollution and extreme heat. In *Planting healthy air: A global analysis of the role of urban trees in addressing particulate matter pollution and extreme heat.*

McEachan, R. R. C., Prady, S. L., Smith, G., Fairley, L., Cabieses, B., Gidlow, C., … & Nieuwenhuijsen, M. J. (2016). The association between green space and depressive symptoms in pregnant women: Moderating roles of socioeconomic status and physical activity. *Journal of Epidemiology and Community Health, 70*(3), 253–259.

McKinney, M. L. (2002). Urbanization, biodiversity, and conservation. *BioScience, 52*, 883–890.

Ming, S., Profous, G., & Dembner, S. (1993). Urban forestry in Beijing. *Unasylva (FAO).*

Mitchell, R., & Popham, F. (2008). Effect of exposure to natural environment on health inequalities: An observational population study. *The Lancet, 372*(9650), 1655–1660.

Mizen, A., Song, J., Fry, R., Akbari, A., Berridge, D., Parker, S. C., … & Rodgers, S. E. (2019). Longitudinal access and exposure to green-blue spaces and individual-level mental health and well-being: Protocol for a longitudinal, population-wide record-linked natural experiment. *British Medical Journal Open, 9*(4), e027289.

Moghadam, H. S., & Helbich, M. (2013). Spatiotemporal urbanization processes in the megacity of Mumbai, India: A Markov chains-cellular automata urban growth model. *Applied Geography, 40*, 140–149. https://doi.org/10.1016/j.apgeog.2013.01.009

Mohan, M., & Kandya, A. (2015). Impact of urbanization and land-use/land-cover change on diurnal temperature range: A case study of tropical urban airshed of India using remote sensing data. *Science of the Total Environment, 506–507*, 453–465. https://doi.org/10.1016/j.scitotenv.2014.11.006

Mõisja, K., Uuemaa, E., & Oja, T. (2016). Integrating small-scale landscape elements into land use/cover: The impact on landscape metrics' values. *Ecological Indicators, 67*, 714–722.

Mondal, A., Mukhopadhyay, A., Guha, S., Kundu, S., Mukherjee, S., & Dasgupta, R. (2012). Decadal-scale vegetation dynamics of Kolkata and its surrounding areas, India using fuzzy classification technique. *Environment and Natural Resources Research, 2*(4), 18.

Mondal, B. K. (2013). Destruction of urban Greenary of Indian cities—A study of the two wards of Kolkata through GIS and remote sensing. *Journal of the Geographical Institute Jovan Cvijic SASA, 63*(4), 93–110.

Morton, F. I. (1983). Operational estimates of areal evapotranspiration and their significance to the science and practice of hydrology. *Journal of Hydrology, 66*(1–4), 1–76. https://doi.org/10.1016/0022-1694(83)90177-4

Moughtin, C., & Shirley, P. (2005). *Urban design: Green dimensions*. Routledge.

Mueller, N., Rojas-Rueda, D., Basagaña, X., Cirach, M., Cole-Hunter, T., Dadvand, P., … & Nieuwenhuijsen, M. (2017). Urban and transport planning related exposures and mortality: A health impact assessment for cities. *Environmental Health Perspectives, 125*(1), 89–96.

Mukherjee, A., Banerjee, A., Kundu, A., & Chatterjee, R. (2018a). Urban green space dynamics and habitat fragmentation: A case study on Kolkata Metropolitan Area, India. *Urban Forestry & Urban Greening, 29*, 26–36.

Mukherjee, S., Bebermeier, W., & Schütt, B. (2018b). An overview of the impacts of land use land cover changes (1980–2014) on urban water security of Kolkata. *Land, 7*(3), 91.

Mukherjee, S., Bebermeier, W., & Schütt, B. (2018c). An overview of the impacts of land use land cover changes (1980–2014) on urban water security of Kolkata. *Land, 7*(3), 91.

Mukhopadhaya, S. (2016). Land use and land cover change modelling using CA-Markov Case study: Deforestation Analysis of Doon Valley. *Journal of Agroecology and Natural Resource Management, 3*, 1–5.

Mushore, T., Mutanga, O., Odindi, J., & Dube, T. (2017). Linking major shifts in land surface temperatures to long term land use and land cover changes: A case of Harare, Zimbabwe. *Urban Climate, 20*, 120–134. https://doi.org/10.1016/j.uclim.2017.04.005

Naikoo, M. W., Rihan, M., Shahfahad, Peer, A. H., Talukdar, S., Mallick, J., ... & Rahman, A. (2022). Analysis of peri-urban land use/land cover change and its drivers using geospatial techniques and geographically weighted regression. *Environmental Science and Pollution Research*, 1–19.

Nath, B., Wang, Z., Ge, Y., Islam, K., P. Singh, R., & Niu, Z. (2020). Land use and land cover change modeling and future potential landscape risk assessment using Markov-CA model and analytical hierarchy process. *ISPRS International Journal of Geo-Information, 9*(2), 134.

Nayak, S., & Mandal, M. (2012). Impact of land-use and land-cover changes on temperature trends over western India. *Current Science, 102*. https://www.jstor.org/stable/24107759.

Neuvonen, M., Sievänen, T., Tönnes, S., & Koskela, T. (2007). Access to green areas and the frequency of visits—A case study in Helsinki. *Urban Forestry & Urban Greening, 6*(4), 235–247.

Newnham, G. J., Verbesselt, J., Grant, I. F., & Anderson, S. A. (2011). Relative Greenness Index for assessing curing of grassland fuel. *Remote Sensing of Environment, 115*(6), 1456–1463.

Nielsen, T. S., & Hansen, K. B. (2007). Do green areas affect health? Results from a Danish survey on the use of green areas and health indicators. *Health & Place, 13*(4), 839–850.

Nieuwenhuijsen, M. J. (2016). Urban and transport planning, environmental exposures and health-new concepts, methods and tools to improve health in cities. *Environmental Health, 15*, 161–171.

Noor, N. M., Abdullah, A., & Manzahani, M. N. H. (2013). Land cover change detection analysis on urban green area loss using GIS and remote sensing techniques. *Planning Malaysia, 11*.

Nourqolipour, R., Shariff, A. R. B. M., Balasundram, S. K., Ahmad, N. B., Sood, A. M., & Buyong, T. (2016). Predicting the effects of urban development on land transition and spatial patterns of land use in Western Peninsular Malaysia. *Applied Spatial Analysis and Policy, 9*(1), 1–19.

Nowak, D. J., Crane, D. E., & Stevens, J. C. (2006). Air pollution removal by urban trees and shrubs in the United States. *Urban Forestry & Urban Greening, 4*(3–4), 115–123.

Nowak, D. J., Noble, M. H., Sisinni, S. M., & Dwyer, J. F. (2001). People and trees: Assessing the US urban forest resource. *Journal of Forestry, 99*(3), 37–42.

Nowak, D. J., Robert III, E., Crane, D. E., Stevens, J. C., & Fisher, C. L. (2010). Assessing urban forest effects and values, Chicago's urban forest. In *Resource Bulletin NRS-37* (pp. 1–27). US Department of Agriculture, Forest Service, Northern Research Station 27.

Nuissl, H., Haase, D., Lanzendorf, M., & Wittmer, H. (2009). Environmental impact assessment of urban land use transitions—A context-sensitive approach. *Land Use Policy, 26*(2), 414–424.

Nutsford, D., Pearson, A. L., & Kingham, S. (2013). An ecological study investigating the association between access to urban green space and mental health. *Public Health, 127*(11), 1005–1011.

Oliveira, S., Andrade, H., & Vaz, T. (2011). The cooling effect of green spaces as a contribution to the mitigation of urban heat: A case study in Lisbon. *Building and Environment, 46*(11), 2186–2194. https://doi.org/10.1016/j.buildenv.2011.04.034

Oliveira, S., Andrade, H., & Vaz, T. (2011b). The cooling effect of green spaces as a contribution to the mitigation of urban heat: A case study in Lisbon. *Building and Environment, 46*(11), 2186–2194.

Oquendo-Di Cosola, V., Olivieri, F., & Ruiz-García, L. (2022). A systematic review of the impact of green walls on urban comfort: Temperature reduction and noise attenuation. *Renewable and Sustainable Energy Reviews, 162*, 112463.

Owen, L. A., Pickering, K. T., & Pickering, K. T. (2006). *An introduction to global environmental issues*. Routledge.

Pandey, B., & Seto, K. C. (2015). Urbanization and agricultural land loss in India: Comparing satellite estimates with census data. *Journal of Environmental Management, 148*, 53–66.

Pandey, D. N., Gupta, A. K., & Ravindranath, N. H. (2010). *Climate change impacts, mitigation and adaptation: Science for generating policy options in Rajasthan, India*. Rajasthan State Pollution Control Board.

Paquet, C., Orschulok, T. P., Coffee, N. T., Howard, N. J., Hugo, G., Taylor, A. W., … & Daniel, M. (2013). Are accessibility and characteristics of public open spaces associated with a better cardiometabolic health?. *Landscape and Urban Planning, 118*, 70–78.

Parihar, S. M., Sarkar, S., Dutta, A., Sharma, S., & Dutta, T. (2013). Characterizing wetland dynamics: A post-classification change detection analysis of the East Kolkata Wetlands using open source satellite data. *Geocarto International, 28*(3), 273–287.

Pauleit, S., Ennos, R., & Golding, Y. (2005). Modeling the environmental impacts of urban land use and land cover change—A study in Merseyside, UK. *Landscape and Urban Planning, 71*(2–4), 295–310.

Pheasant, R. J., Horoshenkov, K. V., & Watts, G. R. (2010). Tranquillity rating prediction tool (TRAPT). *Acoustics Bulletin, 35*(6), 18–24.

Pinho, P., Bergamini, A., Carvalho, P., Branquinho, C., Stofer, S., Scheidegger, C., & Máguas, C. (2012). Lichen functional groups as ecological indicators of the effects of land-use in Mediterranean ecosystems. *Ecological Indicators, 15*(1), 36–42.

Prenzel, B. (2004). Remote sensing-based quantification of land-cover and land-use change for planning. *Progress in Planning, 4*(61), 281–299.

Prüss-Üstün, A., Wolf, J., Corvalán, C., Bos, R., & Neira, M. (2016). *Preventing disease through healthy environments: A global assessment of the burden of disease from environmental risks*. World Health Organization.

Public Works Department (PWD) Kolkata.

Putra, I. G. N. E., Astell-Burt, T., Cliff, D. P., Vella, S. A., John, E. E., & Feng, X. (2020). The relationship between green space and prosocial behaviour among children and adolescents: A systematic review. *Frontiers in Psychology, 11*, 859.

Puyravaud, J. P. (2003). Standardizing the calculation of the annual rate of deforestation. *Forest Ecology and Management, 177*(1–3), 593–596.

Qiu, Y., Zuo, S., Yu, Z., Zhan, Y., & Ren, Y. (2021). Discovering the effects of integrated green space air regulation on human health: A bibliometric and meta-analysis. *Ecological Indicators, 132*, 108292.

Rafiee, R., Mahiny, A. S., & Khorasani, N. (2009). Assessment of changes in urban green spaces of Mashad city using satellite data. *International Journal of Applied Earth Observation and Geoinformation, 11*(6), 431–438.

Rahaman, M., Dutta, S., Sahana, M., & Das, D. N. (2019). Analysing urban sprawl and spatial expansion of Kolkata Urban Agglomeration using geospatial approach. In *Applications and challenges of geospatial technology* (pp. 205–221). Springer.

Rahman, M. M., & Szabó, G. (2021). Impact of land use and land cover changes on urban ecosystem service value in Dhaka, Bangladesh. *Land, 10*(8), 793.

Rai, P. K. (2016). Impacts of particulate matter pollution on plants: Implications for environmental biomonitoring. *Ecotoxicology and Environmental Safety, 129*, 120–136. https://doi.org/10.1016/j.ecoenvol2016.03.012

Rajib, S. Role of urban forestry towards creation of ecocities a case study of Kolkata.

Rakhshandehroo, M., Mohd Yusof, M. J., Arabi, R., Parva, M., & Nochian, A. (2017). The environmental benefits of urban open green spaces. *Alam Cipta, 10*(1), 10–16.

Ramachandra, T. V., Aithal, B. H., & Sreekantha, S. (2012). Spatial metrics based landscape structure and dynamics assessment for an emerging Indian megalopolis. *Facilities, 1*(1).

Ramon, M., Ribeiro, A. P., Theophilo, C. Y. S., Moreira, E. G., de Camargo, P. B., de Bragança Pereira, C. A., … & Ferreira, M. L. (2023). Assessment of four urban forest as environmental indicator of air quality: A study in a brazilian megacity. *Urban Ecosystems, 26*(1), 197–207.

Ray, R., Bandyopadhyay, S., Bhadra, A., Ghosh, S., & Roy, M. (2021). Waterlogging vulnerability assessment of Kolkata City, India: An analysis of urbanization and climate change factors. *Journal of Hydrology, 597*, 126073.

Reveshty, M. A. (2011). The assessment and predicting of land use changes to urban area using multi-temporal satellite imagery and GIS: A case study on Zanjan, IRAN (1984–2011). *Journal of Geographic Information System, 3*(4), 298.

Ring, Z., Damyanovic, D., & Reinwald, F. (2021). Green and open space factor Vienna: A steering and evaluation tool for urban green infrastructure. *Urban Forestry & Urban Greening, 62*, 127131.

Rinner, C., & Hussain, M. (2011). Toronto's urban heat island-Exploring the relationship between land use and surface temperature. *Remote Sensing, 3*(6), 1251–1265. https://doi.org/10.3390/rs3061251

Roe, J. J., Thompson, C. W., Aspinall, P. A., Brewer, M. J., Duff, E. I., Miller, D., … & Clow, A. (2013a). Green space and stress: Evidence from cortisol measures in deprived urban communities. *International Journal of Environmental Research and Public Health, 10*(9), 4086–4103.

Roe, J. J., Thompson, C. W., Aspinall, P. A., Brewer, M. J., Duff, E. I., Miller, D., … & Clow, A. (2013b). Green space and stress: Evidence from cortisol measures in deprived urban communities. *International Journal of Environmental Research and Public Health, 10*(9), 4086–4103.

Rosenberg, N. J., Blad, B. L., & Verma, S. B. (1983). *Microclimate: The biological environment*. Wiley.

Rosenzweig, C., Solecki, W., & Slosberg, R. (2006). Mitigating New York City's heat island with urban forestry, living roofs, and light surfaces. In *A report to the New York state energy research and development authority*.

Rotem-Mindali, O., Michael, Y., Helman, D., & Lensky, I. M. (2015). The role of local land-use on the urban heat island effect of Tel Aviv as assessed from satellite remote sensing. *Applied Geography, 56*, 145–153. https://doi.org/10.1016/j.apgeog.2014.11.023

Roy, M. (2009). Planning for sustainable urbanisation in fast growing cities: Mitigation and adaptation issues addressed in Dhaka, Bangladesh. *Habitat International, 33*(3), 276–286.

Roy, M., Chakraborty, A., & Chakraborty, T. (2020). Evaluating the effectiveness of public awareness campaigns on urban green spaces in Kolkata, India. *Journal of Environmental Planning and Management, 63*(9), 1680–1696.

Roy, P. S., Meiyappan, P., Joshi, P. K., Kale, M. P., Srivastav, V. K., Srivasatava, S. K., ... & Krishnamurthy, Y. V. N. (2016). *Decadal land use and land cover classifications across India, 1985, 1995, 2005*. ORNL DAAC.

Rumelhart, D. E., Hinton, G. E., William, R. J. (1986). Learning internal representation by back-propagation errors. In: Rumelhart, D. E., McClelland, J. L., & PDP Chapter Group (Eds.), Parallel distributed processing: Explorations in microstructure of eCognition. MIT Press.

Sahana, M., & Sajjad, H. (2019). Assessing influence of erosion and accretion on landscape diversity in Sundarban Biosphere Reserve, Lower Ganga Basin: a geospatial approach. In *Quaternary geomorphology in India* (pp. 191–203). Springer.

Sahana, M., Hong, H., & Sajjad, H. (2018). Analyzing urban spatial patterns and trend of urban growth using urban sprawl matrix: A study on Kolkata urban agglomeration, India. *Science of the Total Environment, 628*, 1557–1566.

Sakieh, Y., Salmanmahiny, A., Mirkarimi, S. H., & Saeidi, S. (2017). Measuring the relationships between landscape aesthetics suitability and spatial patterns of urbanized lands: An informed modelling framework for developing urban growth scenarios. *Geocarto International, 32*(8), 853–873.

Salman, M. A. (2004). *A modelling approach to cumulative effects assessment for rehabilitation of remnant vegetation*. PhD Thesis. SRES, The Australian National University.

Salomons, E. M., & Pont, M. B. (2012). Urban traffic noise and the relation to urban density, form, and traffic elasticity. *Landscape and Urban Planning, 108*(1), 2–16.

Sannigrahi, S., Bhatt, S., Rahmat, S., Uniyal, B., Banerjee, S., Chakraborti, S., et al. (2018). Analyzing the role of biophysical compositions in minimizing urban land surface temperature and urban heating. *Urban Climate, 24*, 803–819. https://doi.org/10.1016/j.uclim.2017.10.002

Santos, T., Tenedório, J. A., & Gonçalves, J. A. (2016). Quantifying the city's green area potential gain using remote sensing data. *Sustainability, 8*(12), 1247.

Sarkar, C., Webster, C., & Gallacher, J. (2018). Residential greenness and prevalence of major depressive disorders: A cross-sectional, observational, associational study of 94 879 adult UK Biobank participants. *The Lancet Planetary Health, 2*(4), e162–e173.

Sato, Y., & Yamamoto, K. (2005). Population concentration, urbanization, and demographic transition. *Journal of Urban Economics, 58*(1), 45–61.

Schipperijn, J., Bentsen, P., Troelsen, J., Toftager, M., & Stigsdotter, U. K. (2013). Associations between physical activity and characteristics of urban green space. *Urban Forestry & Urban Greening, 12*(1), 109–116.

Sertel, E., Ormeci, C., & Robock, A. (2011). Modelling land cover change impact on the summer climate of the Marmara region. *Turkey. International Journal of Global Warming, 3*(1/2), 194.

Shaddick, G., Thomas, M. L., Mudu, P., Ruggeri, G., & Gumy, S. (2020). Half the world's population are exposed to increasing air pollution. *NPJ Climate and Atmospheric Science, 3*(1), 23.

Shahfahad, Bindajam, A. A., Naikoo, M. W., Horo, J. P., Mallick, J., Rihan, M., ... & Rahman, A. (2023). Response of soil moisture and vegetation conditions in seasonal variation of land surface temperature and surface urban heat island intensity in sub-tropical semi-arid cities. *Theoretical and Applied Climatology*, 1–29.

Shahfahad, Naikoo, M. W., Das, T., Talukdar, S., Asgher, M. S., Asif, & Rahman, A. (2022). Prediction of land use changes at a metropolitan city using integrated cellular automata: past and future. *Geology, Ecology, and Landscapes*, 1–19.

Shahfahad, Talukdar, S., Naikoo, M. W., Rahman, A., Gagnon, A. S., Islam, A. R. M. T., & Mosavi, A. (2023). Comparative evaluation of Operational Land Imager sensor on board Landsat 8 and Landsat 9 for land use land cover mapping over a heterogeneous landscape. *Geocarto International, 38*(1), 2152496.

Shahfahad, Talukdar, S., Rihan, M., Hang, H. T., Bhaskaran, S., & Rahman, A. (2021). Modelling urban heat island (UHI) and thermal field variation and their relationship with land use indices over Delhi and Mumbai metro cities. *Environment, Development and Sustainability*, 1–29.

Shanahan, D. F., Fuller, R. A., Bush, R., Lin, B. B., & Gaston, K. J. (2015). The health benefits of urban nature: How much do we need? *BioScience, 65*(5), 476–485.

Siddique, G., Roy, A., Mandal, M. H., Ghosh, S., Basak, A., Singh, M., & Mukherjee, N. (2020). An assessment on the changing status of urban green space in Asansol city, West Bengal. *GeoJournal,* 1–23.

Sieghardt, M., Mursch-Radlgruber, E., Paoletti, E., Couenberg, E., Dimitrakopoulus, A., Rego, F., et al. (2005). The abiotic urban environment: Impact of urban growing conditions on urban vegetation. In C. Konijnendijk, K. Nilsson, T. Randrup, & J. Schipperijn (Eds.), *Urban forests and trees* (pp. 281–323). Springer.

Silva, L. T., Oliveira, M., & Silva, J. F. (2014). Urban form indicators as proxy on the noise exposure of buildings. *Applied Acoustics, 76*, 366–376.

Singh, V. S., Pandey, D. N., & Chaudhry, P. (2010). *Urban forests and open green spaces: Lessons for Jaipur, Rajasthan India* (Vol. 1). Rajasthan State Pollution Control Board.

Sivaramakrishnan, K. C., Kundu, A., & Singh, B. N. (2005). *Handbook of urbanization in India: An analysis of trends and processes.*

Slätmo, E., Nilsson, K., & Turunen, E. (2019). Implementing green infrastructure in spatial planning in Europe. *Land, 8*(4), 62.

Smith, G. M., Spencer, T., Murray, A. L., & French, J. R. (1998). Assessing seasonal vegetation change in coastal wetlands with airborne remote sensing: An outline methodology. *Mangroves and Salt Marshes, 2*(1), 15–28.

Smith, N., Georgiou, M., King, A. C., Tieges, Z., Webb, S., & Chastin, S. (2021). Urban blue spaces and human health: A systematic review and meta-analysis of quantitative studies. *Cities, 119*, 103413.

Sorensen, M., Smit, J., Barzetti, V., & Williams, J. (1997). Good practices for urban greening. *Inter-American Development Bank.*

Sturm, R., & Cohen, D. (2014). Proximity to urban parks and mental health. *The Journal of Mental Health Policy and Economics, 17*(1), 19.

Su, S., Xiao, R., Jiang, Z., & Zhang, Y. (2012). Characterizing landscape pattern and ecosystem service value changes for urbanization impacts at an eco-regional scale. *Applied Geography, 34*, 295–305.

Sudha, P., & Ravindranath, N. H. (2000). A study of Bangalore urban forest. *Landscape and Urban Planning, 47*(1–2), 47–63.

Sugiyama, T., Villanueva, K., Knuiman, M., Francis, J., Foster, S., Wood, L., & Giles-Corti, B. (2016). Can neighborhood green space mitigate health inequalities? A study of socio-economic status and mental health. *Health & Place, 38*, 16–21.

Sun, C., Lin, T., Zhao, Q., Li, X., Ye, H., Zhang, G., … & Zhao, Y. (2019). Spatial pattern of urban green spaces in a long-term compact urbanization process—A case study in China. *Ecological Indicators, 96*, 111–119.

Sun, X., Tan, X., Chen, K., Song, S., Zhu, X., & Hou, D. (2020). Quantifying landscape-metrics impacts on urban green-spaces and water-bodies cooling effect: The study of Nanjing, China. *Urban Forestry & Urban Greening, 55*, 126838.

Sun, Z., & Zhu, D. (2019). Exposure to outdoor air pollution and its human-related health outcomes: An evidence gap map. *British Medical Journal Open, 9*(12), e031312.

Tallis, M. J., Amorim, J. H., Calfapietra, C., Freer-Smith, P., Kotthaus, S., Toscano, P., ... & Grimmond, S. (2015). The impacts of green infrastructure on air quality and temperature. In *Handbook on green infrastructure* (pp. 30–49). Edward Elgar Publishing.

Tang, U. W., & Wang, Z. S. (2007). Influences of urban forms on traffic-induced noise and air pollution: Results from a modelling system. *Environmental Modelling & Software, 22*(12), 1750–1764.

Taubenböck, H., Wegmann, M., Roth, A., Mehl, H., & Dech, S. (2009). Urbanization in India-Spatiotemporal analysis using remote sensing data. *Computers, Environment and Urban Systems, 33*(3), 179–188.

Taylor, L., & Hochuli, D. F. (2017). Defining greenspace: Multiple uses across multiple disciplines. *Landscape and Urban Planning, 158*, 25–38.

Teotónio, I., Silva, C. M., & Cruz, C. O. (2021). Economics of green roofs and green walls: A literature review. *Sustainable Cities and Society, 69*, 102781.

The World Resources Institute. (1996). *The urban environment: World resources 1996–97*. Oxford University Press.

Thurston, G. D., Kipen, H., Annesi-Maesano, I., Balmes, J., Brook, R. D., Cromar, K., ... & Brunekreef, B. (2017). A joint ERS/ATS policy statement: What constitutes an adverse health effect of air pollution? An analytical framework. *European Respiratory Journal, 49*(1).

Tok, E., Agdas, M. G., Ozkok, M. K., & Kuru, A. (2020). Socio-psychological effects of urban green areas: Case of Kirklareli City Center. *Journal of Contemporary Urban Affairs, 4*(1), 47–60.

Townshend, J. R. G., & Justice, C. O. (1995). Spatial variability of images and the monitoring of changes in the normalized difference vegetation index. *International Journal of Remote Sensing, 16*(12), 2187–2195.

Tzoulas, K., Korpela, K., Venn, S., Yli-Pelkonen, V., Kaźmierczak, A., Niemela, J., & James, P. (2007). Promoting ecosystem and human health in urban areas using Green Infrastructure: A literature review. *Landscape and Urban Planning, 81*(3), 167–178.

Ulrich, R. S. (1984). View through a window may influence recovery from surgery. *Science, 224*(4647), 420–421.

UN, United Nations. (2014). *World urbanization prospects World urbanization prospects: The 2014 revision highlights*. https://doi.org/10.4054/DemRes.2005.12.9

United Nations Department of Economic and Social Afairs (UNDESA). (2012). *World urbanization prospects: The 2011 revision*. United Nations Department of Economic and Social Afairs/Population Division.

United Nations. (2012). *World urbanization prospects: The 2011 revision*. United Nations.

Uy, P. D., & Nakagoshi, N. (2007). Analyzing urban green space pattern and eco-network in Hanoi, Vietnam. *Landscape and Ecological Engineering, 3*, 143–157.

Valente-Pereira, L. (2014). *Urban form definition in urban planning*. Simplíssimo.

Van den Berg, A. E., Maas, J., Verheij, R. A., & Groenewegen, P. P. (2010). Green space as a buffer between stressful life events and health. *Social Science & Medicine, 70*(8), 1203–1210.

Van den Bosch, M., & Sang, Å. O. (2017). Urban natural environments as nature-based solutions for improved public health—A systematic review of reviews. *Environmental Research, 158*, 373–384.

Van Herzele, A., & Wiedemann, T. (2003). A monitoring tool for the provision of accessible and attractive urban green spaces. *Landscape and Urban Planning, 63*(2), 109–126.

Van Renterghem, T., Attenborough, K., Maennel, M., Defrance, J., Horoshenkov, K., Kang, J., … & Yang, H. S. (2014). Measured light vehicle noise reduction by hedges. *Applied Acoustics, 78*, 19–27.

Van Renterghem, T., Botteldooren, D., & Verheyen, K. (2012). Road traffic noise shielding by vegetation belts of limited depth. *Journal of Sound and Vibration, 331*(10), 2404–2425.

Vanaken, G. J., & Danckaerts, M. (2018). Impact of green space exposure on children's and adolescents' mental health: A systematic review. *International Journal of Environmental Research and Public Health, 15*(12), 2668.

Velmurugan, A., & Sajjad, H. (2009). The study of land transformation and land degradation in Dehradun District, Uttrakhand. *Deccan Geographer, 48*.

Viecco, M., Jorquera, H., Sharma, A., Bustamante, W., Fernando, H. J., & Vera, S. (2021). Green roofs and green walls layouts for improved urban air quality by mitigating particulate matter. *Building and Environment, 204*, 108120.

Vienneau, D., de Hoogh, K., Faeh, D., Kaufmann, M., Wunderli, J. M., Röösli, M., & SNC Study Group. (2017). More than clean air and tranquillity: Residential green is independently associated with decreasing mortality. *Environment International, 108*, 176–184.

Vijayaraghavan, K. (2016). Green roofs: A critical review on the role of components, benefits, limitations and trends. *Renewable and Sustainable Energy Reviews, 57*, 740–752.

Votsi, N. E. P., Drakou, E. G., Mazaris, A. D., Kallimanis, A. S., & Pantis, J. D. (2012). Distance-based assessment of open country Quiet Areas in Greece. *Landscape and Urban Planning, 104*(2), 279–288.

Vymazal, J. (2007). Removal of nutrients in various types of constructed wetlands. *Science of the Total Environment, 380*(1–3), 48–65.

Wan, C., Shen, G. Q., & Choi, S. (2021). Underlying relationships between public urban green spaces and social cohesion: A systematic literature review. *City, Culture and Society, 24*, 100383.

Wang, J., Zhou, W., Qian, Y., Li, W., & Han, L. (2018). Quantifying and characterizing the dynamics of urban greenspace at the patch level: A new approach using object-based image analysis. *Remote Sensing of Environment, 204*, 94–108.

Wang, X. J. (2009). Analysis of problems in urban green space system planning in China. *Journal of Forestry Research, 20*(1), 79–82.

Ward, D. S., Mahowald, N. M., & Kloster, S. (2014). Potential climate forcing of land use and land cover change. *Atmospheric Chemistry and Physics, 14*(23), 12701–12724.

Weber, N., Haase, D., & Franck, U. (2014). Traffic-induced noise levels in residential urban structures using landscape metrics as indicators. *Ecological Indicators, 45*, 611–621.

Weng, Q. (2001). A remote sensing-GIS evaluation of urban expansion and its impact on surface temperature in Zhujiang Delta, China. *International Journal of Remote Sensing, 22*(10), 1999–2014.

Weng, Q., Lu, D., & Schubring, J. (2004). Estimation of land surface temperature-vegetation abundance relationship for urban heat island studies. *Remote Sensing of Environment, 89*(4), 467–483.

WHO. (2010d). *Urban planning, environment and health: From evidence to policy action. Meeting report*. WHO Regional Office for Europe.

Wolch, J. R., Byrne, J., & Newell, J. P. (2014a). Urban green space, public health, and environmental justice: The challenge of making cities 'just green enough.' *Landscape and Urban Planning, 125*, 234–244.

Wolch, J. R., Byrne, J., & Newell, J. P. (2014b). Urban green space, public health, and environmental justice: The challenge of making cities 'just green enough.' *Landscape and Urban Planning, 125*, 234–244. https://doi.org/10.1016/j.landurbplan.2014.01.017

Wolf, K. L., Lam, S. T., McKeen, J. K., Richardson, G. R., Van Den Bosch, M., & Bardekjian, A. C. (2020). Urban trees and human health: A scoping review. *International Journal of Environmental Research and Public Health, 17*(12), 4371.

World Health Organization. (2013). *Physical activity promotion in socially disadvantaged groups: principles for action: PHAN Work Package 4* (No. WHO/EURO: 2013-4480-44243-62498). World Health Organization. Regional Office for Europe.

World Health Organization. (2016). *World Health Statistics 2016 [OP]: Monitoring Health for the Sustainable Development Goals (SDGs)*. World Health Organization.

World Health Organization. (2018). Physical activity for health: More active people for a healthier world: Draft global action plan on physical activity 2018–2030. In *Seventy-First World Health Assembly Provisional Agenda Item 12.2.* Retrieved from: http://apps.who.int/gb/ebwha/pdf_files/WHA71/A71_18-en.pdf

Wu, H., Yang, C., Chen, J., Yang, S., Lu, T., & Lin, X. (2018). Effects of Green space landscape patterns on particulate matter in Zhejiang Province, China. *Atmospheric Pollution Research, 9*(5), 923–933.

Wu, J. (2014). Urban ecology and sustainability: The state-of-the-science and future directions. *Landscape and Urban Planning, 125*, 209–221.

Wu, J., Wang, X., Zhong, B., Yang, A., Jue, K., Wu, J., … & Liu, Q. (2020). Ecological environment assessment for Greater Mekong Subregion based on pressure-state-response framework by remote sensing. *Ecological Indicators, 117*, 106521.

Wu, J., Yang, M., Xiong, L., Wang, C., & Ta, N. (2021). Health-oriented vegetation community design: Innovation in urban green space to support respiratory health. *Landscape and Urban Planning, 205*, 103973.

Wu, K. Y., & Zhang, H. (2012). Land use dynamics, built-up land expansion patterns, and driving forces analysis of the fast-growing Hangzhou metropolitan area, eastern China (1978–2008). *Applied Geography, 34*, 137–145.

Wu, Z., Chen, R., Meadows, M. E., Sengupta, D., & Xu, D. (2019). Changing urban green spaces in Shanghai: Trends, drivers and policy implications. *Land Use Policy, 87*, 104080.

Xian, C., Gong, C., Wu, T., Liu, J., & Ouyang, Z. (2022). *Air pollutant removal by urban vegetation: A meta-analysis of gaps in research and environmental management.*

Xian, G., Crane, M., & Steinward, D. (2005). Dynamic modeling of Tampa Bay urban development using parallel computing. *Computers & Geosciences, 31*(7), 920–928. https://doi.org/10.1016/j.cageo.2005.03.006

Xiao, R., Weng, Q., Ouyang, Z., Li, W., Schienke, E. W., & Zhang, Z. (2008). Land surface temperature variation and major factors in Beijing, China. *Photogrammetric Engineering & Remote Sensing, 74*(4), 451–461. https://doi.org/10.14358/PERS.74.4.451

Xu, C., Haase, D., & Pauleit, S. (2018). The impact of different urban dynamics on green space availability: A multiple scenario modeling approach for the region of Munich, Germany. *Ecological Indicators, 93*, 1–12.

Xu, L. Y., Xie, X. D., & Li, S. (2013). Correlation analysis of the urban heat island effect and the spatial and temporal distribution of atmospheric particulates using TM images in Beijing. *Environmental Pollution, 178*, 102–114. https://doi.org/10.1016/j.envpol.2013.03.006

Yan, J., Lin, L., Zhou, W., Han, L., & Ma, K. (2016). Quantifying the characteristics of particulate matters captured by urban plants using an automatic approach. *Journal of Environmental Sciences, 39*, 259–267.

Yang, W., & Jeon, J. Y. (2020). Design strategies and elements of building envelope for urban acoustic environment. *Building and Environment, 182*, 107121.

Yogesh, K., Bharath, B. D., Mallick, J., Atzberger, C., & Kerle, N. (2009). Satellite-based analysis of the role of land use: Land cover and vegetation density on surface temperature regime of Delhi, India. *Journal of the Indian Society of Remote Sensing, 37*(2), 201–214. https://doi.org/10.1007/s12524-009-0030-x

You, G., Zhang, Y., Liu, Y., Schaefer, D., Gong, H., Gao, J., Lu, Z., Song, Q., Zhao, J., Wu, C., Yu, L., & Xie, Y. (2013). Investigation of temperature and aridity at different elevations of Mt. Ailao, SW China. *International Journal of Biometeorology, 57*(3), 487–492.

Yu, B., Bai, J., Wen, L., & Chai, Y. (2022). Psychophysiological impacts of traffic sounds in urban green spaces. *Forests, 13*(6), 960.

Yuan, F., & Bauer, M. E. (2007). Comparison of impervious surface area and normalized difference vegetation index as indicators of surface urban heat island effects in Landsat imagery. *Remote Sensing of Environment, 106*(3), 375–386.

Yuan, F., Sawaya, K. E., Leofelholz, C. B., & Bauer, M. E. (2005a). Land cover classification and change analysis of the twin cities (Minnesota) metropolitan area by multi-temporal Landsat remote sensing. *Remote Sensing of Environment, 98*, 317–328. https://doi.org/10.1016/j.rse.2005.08.006

Yuan, F., Sawaya, K. E., Loeffelholz, B. C., & Bauer, M. E. (2005b). Land cover classification and change analysis of the Twin Cities (Minnesota) Metropolitan Area by multitemporal Landsat remote sensing. *Remote Sensing of Environment, 98*(2–3), 317–328.

Yuan, F., Sawaya, K. E., Loeffelholz, B. C., & Bauer, M. E. (2005c). Land cover classification and change analysis of the Twin Cities (Minnesota) metropolitan area by multitemporal landsat remote sensing. *Remote Sensing of Environment, 98*(2–3), 317–328.

Zha, Y., Gao, J., & Ni, S. (2003). Use of normalized difference built-up index in automatically mapping urban areas from TM imagery. *International Journal of Remote Sensing, 24*(3), 583–594.

Zhang, B., Xie, G. D., Gao, J. X., & Yang, Y. (2014). The cooling effect of urban green spaces as a contribution to energy-saving and emission-reduction: A case study in Beijing China. *Building and Environment*. https://doi.org/10.1016/j.buildenv.2014.03.00

Zhang, D., Ling, G. H. T., Misnan, S. H. B., & Fang, M. (2023). A systematic review of factors influencing the vitality of public open spaces: A novel perspective using social-ecological model (SEM). *Sustainability, 15*(6), 5235.

Zhang, F., Xu, N., Wang, C., Wu, F., & Chu, X. (2020). Effects of land use and land cover change on carbon sequestration and adaptive management in Shanghai, China. *Physics and Chemistry of the Earth, Parts a/b/c, 120*, 102948.

Zhang, K. H., & Shunfeng, S. O. N. G. (2003). Rural–urban migration and urbanization in China: Evidence from time-series and cross-section analyses. *China Economic Review, 14*(4), 386–400.

Zhang, R., Zhang, C. Q., & Rhodes, R. E. (2021). The pathways linking objectively-measured greenspace exposure and mental health: A systematic review of observational studies. *Environmental Research, 198*, 111233.

Zhao, J., Chen, S., Jiang, B., Ren, Y., Wang, H., Vause, J., & Yu, H. (2013). Temporal trend of green space coverage in China and its relationship with urbanization over the last two decades. *Science of the Total Environment, 442*, 455–465.

Zhou, F., Xu, Y. P., & Lv, H. H. (2012). Analysis of the vegetation change in the Yangtze River Delta based on Modis-EVI Time Series data. *Resources and Environment in the Yangtze Basin, 21*(11), 1363–1369.

Zhou, W., Troy, A., & Grove, M. (2008). Object-based land cover classification and change analysis in the Baltimore metropolitan area using multitemporal high resolution remote sensing data. *Sensors, 8*(3), 1613–1636.

Zhou, X., & Wang, Y. C. (2011). Spatial–temporal dynamics of urban green space in response to rapid urbanization and greening policies. *Landscape and Urban Planning, 100*(3), 268–277.

Zhu, X., Gao, M., Zhang, R., & Zhang, B. (2021). Quantifying emotional differences in urban green spaces extracted from photos on social networking sites: A study of 34 parks in three cities in northern China. *Urban Forestry & Urban Greening, 62*, 127133.

Index

© The Editor(s) (if applicable) and The Author(s), under exclusive license to Springer Nature Switzerland AG 2026

M. B. Ali and S. Jamal, *Urban Green Spaces and Environmental Sustainability*, https://doi.org/10.1007/978-3-032-15322-7

MIX
Papier aus verantwortungsvollen Quellen
Paper from responsible sources
FSC® C105338

If you have any concerns about our products,
you can contact us on
ProductSafety@springernature.com

In case Publisher is established outside the EU,
the EU authorized representative is:
Springer Nature Customer Service Center GmbH
Europaplatz 3, 69115 Heidelberg, Germany

Printed by Libri Plureos GmbH
in Hamburg, Germany